The Concept of Milieu in Environmental Ethics

The Concept of Milieu in Environmental Ethics discusses how we can come together to address current environmental problems at the planetary level, such as climate change, biodiversity loss, transborder pollution and desertification.

The book recognizes the embedded individual sociocultural and environmental contexts that impact our everyday choices. It asks, in this pluralism of worldviews, how can we build common ground to tackle environmental issues? What is our individual moral responsibility within the larger collaborative challenge? Through philosophical reasoning, this book pragmatically addresses these questions and builds a framework to support sustainable ways of living. At the core of the book, it draws on the concept of milieu (fūdo), inspired by the Japanese philosopher Watsuji Tetsurō, which captures how we act within and perceive our surroundings as a web of culturally, historically and geographically situated meanings and values. It argues that the milieu connects us as individuals with community, past and future history, and the natural world, providing us with common ground for global environmental ethics.

This book will be an engaging and interesting read for scholars, researchers and students in environmental ethics, philosophy and sustainability.

Laÿna Droz recently completed her PhD in Global Environmental Studies at Kyoto University, Japan.

Routledge Environmental Ethics
Series Editor: Benjamin Hale, University of Colorado, Boulder

The Routledge Environmental Ethics series aims to gather novel work on questions that fall at the intersection of the normative and the practical, with an eye toward conceptual issues that bear on environmental policy and environmental science. Recognizing the growing need for input from academic philosophers and political theorists in the broader environmental discourse, but also acknowledging that moral responsibilities for environmental alteration cannot be understood without rooting themselves in the practical and descriptive details, this series aims to unify contributions from within the environmental literature.

Books in this series can cover topics in a range of environmental contexts, including individual responsibility for climate change, conceptual matters affecting climate policy, the moral underpinnings of endangered species protection, complications facing wildlife management, the nature of extinction, the ethics of reintroduction and assisted migration, and reparative responsibilities to restore, among many others.

Philosophy in the American West: A Geography of Thought
Edited by Josh Hates, Gerard Kuperus and Brian Treanor

Philosophy and the Climate Crisis
How the Past can Save the Present
Byron Williston

Climate Justice Beyond the State
Lachlan Umbers and Jeremy Moss

The Concept of Milieu in Environmental Ethics
Individual Responsibility within an Interconnected World
Laÿna Droz

For more information on the series, please visit: https://www.routledge.com/Routledge-Environmental-Ethics/book-series/ENVE

The Concept of Milieu in Environmental Ethics

Individual Responsibility within an Interconnected World

Laÿna Droz

Routledge
Taylor & Francis Group

LONDON AND NEW YORK

First published 2022
by Routledge
2 Park Square, Milton Park, Abingdon, Oxon OX14 4RN

and by Routledge
605 Third Avenue, New York, NY 10158

Routledge is an imprint of the Taylor & Francis Group, an informa business

British Library Cataloguing-in-Publication Data
A catalogue record for this book is available from the British Library

Library of Congress Cataloging-in-Publication Data
Names: Droz, Laÿna, author.
Title: The concept of milieu in environmental ethics : individual
responsibility within an interconnected world / Laÿna Droz.
Description: Milton Park, Abingdon, Oxon ; New York, NY :
Routledge, 2022. | Includes bibliographical references and index.
Identifiers: LCCN 2021007118 (print) | LCCN 2021007119
(ebook) | ISBN 9780367776435 (hardback) | ISBN 9780367776466
(paperback) | ISBN 9781003172208 (ebook)
Subjects: LCSH: Environmental ethics.
Classification: LCC GE42 .D76 2022 (print) | LCC GE42 (ebook) |
DDC 179/.1—dc23
LC record available at https://lccn.loc.gov/2021007118
LC ebook record available at https://lccn.loc.gov/2021007119

ISBN: 978-0-367-77643-5 (hbk)
ISBN: 978-0-367-77646-6 (pbk)
ISBN: 978-1-003-17220-8 (ebk)

Typeset in Bembo
by codeMantra

To my inspiration, my late grandmother Bluette, because I owe it all to you.

Contents

Figures

Tables

Foreword

Yasuo Deguchi

This is a book of philosophical bricolage. Philosophical bricolage aims to create a new worldview or value system through the mixture or crossbreeding, rather than mere comparison or juxtaposition, of different intellectual traditions. Such philosophy often revolves around an idea that has undergone metamorphoses time and again while being relayed from one socio-cultural context to another. In this book, that idea is *milieu*.

The philosophical origins of *milieu* begin with an ancient Chinese word, *fēdŭ,* that literally means 'wind-soil'. *Fēdŭ* was imported to Japan, where it was repronounced as *fūdo.* The both words can be roughly translated as *terroir,* i.e., an integration of the natural and cultural factors—such as climate, soil, and manner of cultivation—that give an agricultural product its particular quality and taste. The Japanese philosopher Tetsurō Watsuji transplanted this idea into the context of German philosopher Martin Heidegger's magnum opus, *Being and Time.* Watsuji rendered its meaning as his own version of the Heideggerian *Dasein,* the human way of being. Watsuji's *fūdo*-as-Dasein still remains mainly—if not exclusively—individualistic, rather than collective. However, French philosopher Augustin Berque translated Watsuji's *fūdo* as *milieu,* articulating the mutual and dynamic relations between human and nature within it and thereby making Watsuji's *fūdo*-as-Dasein more explicitly collective and social.

This book by Laÿna Droz represents the next movement of the cross-cultural philosophical bricolage of *milieu*. Droz is dissatisfied with both Watsuji's *fūdo* and Berque's *milieu,* claiming that they fail to secure the moral and political responsibilities of the individual human for global environmental issues. That is why she enlarges the concept of *milieu* to incorporate both collectives and individuals as two distinct factors within its structure. By so doing, she sharpens Watsuji's *fūdo* and Berque's *milieu* and gives them a practical orientation that the originals lack.

It is still debatable whether Watsuji can be counted among the philosophers of the Kyoto School, the philosophical movement founded by Japanese philosopher Kitarō Nishida. But it seems to me that Watsuji's idea of *fūdo*-as-Dasein conveys the flavor of the school. This would be in accordance with biographical anecdotes. His idea of *fūdo* was incubated in his lectures at Kyoto

University during 1928-29 before he moved to the University of Tokyo in 1934. During his years as a Kyoto professor, Watsuji regularly attended Nishida's lectures, taking notes earnestly at the frontmost bench of the theatre—or so I was told by my grandfather, who attended their lectures in 1920s.

But there are also philosophical reasons to take Watsuji's *fūdo* as belonging to the Kyoto School. The Kyoto School consists of philosophers who tried to reanimate an East Asian traditional view of self, that is, the *true self,* a holistic, embodied, and non-dual (or trans-dichotomous) concept of self. In this way, they tried to overcome the self-centeredness of the individual, while restoring its moral and socio-political responsibilities. Watsuji's *fūdo*-as-Dasein can be taken as a continuation of this project. Watsuji also endeavored to rebuild social norms based on his concept of *fūdo*-as-Dasein, even though Droz and others can rightfully dismiss his efforts as unsatisfactory.

This identity with the Kyoto School extends to Droz herself. Take her idea of self as a restless pendulum that always swings between the self-negation of its independence from the milieu and its self-reappropriation. This too can be seen as an echo of the Kyoto School's notion of *true self.* It is thus appropriate to take this book as a new, environmentally conscious development of Kyoto School philosophy that seriously tackles the global environmental crisis. Let us welcome Laÿna Droz as a brilliant and enterprising new member of the school.

Yasuo Deguchi
Professor of Philosophy, Graduate School of Letters
Vice Provost of Kyoto University

Acknowledgements

My deepest gratitude goes to the galaxy of professors at Kyoto University and beyond who guided me through these formative years, especially to Usami Makoto, Deguchi Yasuo, Marc-Henri Deroche, Kai Yuan-Cheng, Iwatani Ayako and Sato Junji. Their support and advice nurtured my research throughout the years.

Discussions with fellow students and friends in Japan and all over the world are the driving force of my thinking, and I could not have developed my ideas without them. In particular, I would like to thank Larissa Alem, Putri Dwinatalis Baeha, Nicolas Besuchet, Leila Chakroun, Alisa Cosel, Kenny Dong, Rika Fajrini, Natasha Handel, Takeshi Hara, Janiel Hazle, Ritika Jain, Komatsubara Orika, Roxane Kuriowiak, Hùng Sông Lam, Bryant Lin, Mégane Matile, Dinita Setyawati, Maximillian Spielberg, Lei Tian, Bich Ha Ngoc Trinh, Chang Che Wei, Maiko Yamamori and Yang Yang. I also want to thank everyone at the Laboratory of Global Environmental Policy and at the Department of Philosophy at Kyoto University, it was unforgettable sharing study groups and getting to know all of you.

I am also deeply thankful to scholars, researchers and activists that I had the chance to meet and exchange with at various conferences and informal events across the globe, for their precious encouragements and their highly helpful advices: Masahiro Morioka and the Japanese network for philosophy of life; the members of the International Society of Environmental Ethics (in particular Simon James, Benjamin Hale, Eugene Hargrove, Rika Tsuji and Ben Johnson); members of the network of the Mind and Life Summer Research Institute (especially Hanne de Jaegher and Kris Acorn); scholars of the intellectual movement *mésologies* (especially Augustin Berque and Yoann Moreau); participants of the World Philosophy Congress in Beijing (in particular Arto Mutanen, Stelios Virvikadis, Chistopher Volparil, Kurt Jax, and Sally Haslanger); the staff and researchers of the Intergovernmental Science-Policy Platform on Biodiversity and Ecosystem Services (IPBES) Secretariat in Bonn; the members of the European Network for Japanese Philosophy (in particular James Heisig, Hans Peter Liederbach and Takeshi Morisato); scholars at the Research Center for Humanity and Nature (in particular Kazuhiko Ota and Christoph Rupprecht); the activists of the Global

Young Greens; the activists of Green Action Japan (especially Aileen Mioko Smith); the activists of the Greens and Young Greens Switzerland; and also Filippo Casati, Jay Garfield, Inukai Yumiko, Jun Otsuka, Graham Priest, as well as Miyata Akihiro and Anton Sevilla. My research also benefited from the comments of the anonymous reviewers of several peer-reviewed journals.

A very special gratitude goes to my father, Yvan Droz who, not content with raising me, also read and discussed the entirety of my writings. Romaric Jannel and Jimmy Fyfe also proofread, with their sharp philosophical eyes and critical thinking, crucial parts of my writing. Etsuro Ishimaru had the bewildering kind patience to slowly read with me *Rinrigaku* during countless hours, and I could not have understood the subtleties of the Japanese text without him. And Yoko Kanazawa taught me more through her love for nature and farming than any book.

I also benefited from the precious financial support of the Japanese Ministry of Education, Culture, Sports, Science and Technology (Monbukagaku-usho, *MEXT*) scholarship, and from punctual funding to attend international conferences by the Educational Unit for Studies on the Connectivity of Hills, Humans and Oceans (CoHHO).

Last but by no means least, I want to thank all my friends and my family for their support, inspiration and encouragements, and for never reproaching me to have stayed so far away from many of them for so long. Thank you.

1   Introduction

Climate change, biodiversity loss, transborder pollution, desertification… Many current environmental problems span at the planetary level and are caused by the ways of life of countless individuals across the globe. While many of us recognize that we need to collaborate to tackle together the global environmental crisis, each of us remains embedded in sociocultural and environmental contexts and makes choices through our personal worldview. In this global context of pluralism of worldviews, how can we build common grounds to take common action and tackle environmental problems? What is our individual moral responsibility in this giant collaborative task? What do we actually want to sustain?

Philosophical and conceptual reasoning is needed to understand the depths of the current global environmental crisis and to provide some sustainable solutions to these challenges, adapted to individual questionings on how to lead a good life in our present interconnected world.[1] In an attempt to make a humble contribution to this immense enterprise, in this book, I argue that the idea of milieu can ground environmental ethics, anchor sustainability as a normative direction and help rooting individual responsibility for environmental harm. While most contributions in environmental ethics work with normative assumptions anchored in their specific sociocultural contexts, this book positions environmental ethical questions in the global context. It dwells on the concept of milieu (*fūdo*) inspired by the Japanese philosopher Watsuji Tetsurō and applies it in the fields of environmental ethics and global political philosophy. This book aims at building a motivational framework that supports sustainable ways of life, while being sensitive to how the environmental crisis can be overwhelming at times.

Before entering into the discussion of the concept of the milieu and the argumentation, I need to set the stage of the reasoning. I first describe the environmental and social situations that compose the current state of environmental urgency. Second, I describe the dominant key ideas underlying and informing the current situation. Third, I discuss the pragmatist approach of my project determined by the need for consensus. Finally, I describe the methodological implications of this approach and my personal standpoint.

1.1 Environmental urgency

First and foremost, it is important to situate my reasoning in the specific socio-political context and environmentally challenged world in which we are living here and now. Indeed, my project is all but detached from these concrete external constraints. It is driven by the need to understand better my relation to and what I can do about the environmental problems tinged by the socio-political contexts. It is why I start by sketching the contemporary environmental situation and the social situation.

1.1.1 Environmental situation

Our era is characterized by environmental problems such as pollution, climate change and severe biodiversity loss. Environmental problems are often intertwined and mutually reinforcing. For example, the rapid deforestation in the Amazon increases the effects of climate change by reducing the Planet's carbon sink, induces desertification in the long run and worsens biodiversity loss by reducing the habitat of many species. Among them, species extinction is particularly disturbing. As Martin Gorke puts it:

> It is *irreversible,* and it indicates that if destruction of nature continues, a point will eventually be reached at which even our own species may cease to exist or at least may forfeit all hopes of leading a *good* life.
> (Emphasis in original; Gorke 2003)

This quote connects directly the severity of the environmental problems with fundamental philosophical questions such as the nature and meaning of human existence, and the normative assessment of what is a *good* life. Environmental problems are not merely questions of facts and of the best technology to be used to improve the material situation. They question directly our ways of life, our choices of life, the meanings and values that we attach to different ways of leading one's life and to human life itself. In fact, environmental problems are *human* problems because our human ways of life are their main cause, and because they threaten and harm our human ways of life and projects. They are our problem, as we are living intertwined in webs of complex constitutive relations with each other, other species and the interconnected world.

Ethics is the area of philosophy dealing with how we *should* lead our lives, and how we *should* behave in the everyday life. Understood in these terms, ethics exists in every human culture at all times. It is rooted in the very fact that we are able to reflect on ourselves, on our actions and on our behaviours. Until the last century, the realm of ethics was largely confined to behaviours and action between individual fellow humans, and to behaviours between and towards groups of humans. On its margin, it sometimes included behaviours towards other elements of our world, such as animals, plants, rocks, or deities, wind and spirits. As guiding norms and rules of how to live, it was largely localized and tightly intertwined with local cultures, social contexts and specific environmental challenges. Even if people moved and ideas were always exchanged all around our unique planet, for most people, questions of ethics tended to be limited to their current actions towards other beings in presence.

Economic globalization and technologies of communication have extended the reach of our daily actions to spaces and times beyond our physically situated standing point. Most of the environmental problems we are facing nowadays are consequences of a global increase in consumption of material goods, which are processed and transported over long distances. This state of affairs is rooted in a mainstream consumerist framework nourished by ideologies such as economic globalization and liberal capitalism. The champions of technical or technological optimism believe and argue that most of these environmental problems are not to be worried about, because they can be counterbalanced by technological intervention. For instance, the adverse effects of climate change could be mitigated by the deliberate and large-scale manipulation of the Earth's climate system, known as geoengineering. Despite their vehemence, it cannot be expected to solve *all* environmental problems by a sudden improvement in technological knowledge. The best example of this impossibility is habitat destruction and species extinction.

Nature-based solutions aim to protect, sustainably manage and restore ecosystems to allow them to provide benefits and support human well-being. They are claimed to be safer and more cost-effective than technological solutions. For example, coastal mangrove forest can be maintained and restored to act as a buffer between land and sea and protect communities from storms and erosion, which is crucial to face the effects of climate change. Human-made substitutes, such as dikes and seawalls, are costlier and do not provide all the advantages of natural mangrove ecosystems. Nature-based solutions slowly gain importance in the international discourses regarding environmental problems. They show that the solutions to our human environmental problems are all but disconnected from our ways of life with, and our ideas of nature.

Another aspect that highlights the interconnectedness of our lives with the natural world is health. At the individual scale, our health depends on the food we eat, the water we drink and the air we breathe, that all largely come from ecosystems and the natural environment. At a regional scale, a healthy local natural environment gives us the freedom to use the space without fear for our health, be it from air pollution, water and soil contamination, etc. We are vulnerable to changes in nature as individuals and as communities. Moreover, ecosystems are interconnected at the planetary level. Organisms – including invasive species, microorganisms and virus – travel across long distances across the globe. Air pollution and plastic pollution move without any consideration for human political borders. And the effects of climate change wreak havoc globally. The same human activities that drive climate change and biodiversity loss also drive an increasing pandemic risk by reducing species habitats and species reservoirs, increasing contacts between humans, livestock and wildlife, and the possibilities for microbes of animal origin to adapt to human hosts (IPBES 2020). Ideas of planetary health emerged from the recognition of the interdependence between environmental health, animal health and human health. In this view, for human lives to flourish, we need a healthy planetary environment, including habitats for diverse species and ecological connectivity between ecosystems.

The direct drivers with the largest global impacts on the natural world are identified to be changes in land and sea use, direct exploitation of organisms, climate change, pollution and invasion of alien species (IPBES 2019). They emerge from human activities, such as the overexploitation of ecosystems and species, land clearing, unsustainable agriculture and fishing, infrastructures expansion and overconsumption. These human activities are shaped by sociocultural, political and economic factors. Ultimately, they are rooted in ways of life that are anchored in worldviews and ideas regarding what a good life is. Therefore, even if technology improves significantly, develops in collaboration with nature and offers the best possible solutions at a given time, relying exclusively on it would ignore that the roots of the problems are in the choices of behaviours informed by our worldviews and by the ethical reflections we make. The latter may be exactly what make us human, and what give subjective meanings to our individual existence. To reflect about it is thus a worthwhile choice.

1.1.2 Social situation

Nowadays more than ever, individual actions are interwoven with webs of social structures and systems. Rare are those who consume exclusively what they produce with their own hands using homemade tools. Cellphones have conquered the most remote communities making them part of the buzzing world web of exchange of information. But as liberating and free as it might look from a sedentary perspective, this web is compartmentalized and organized by the intervention of states and various other actors and limitations, from the language used to the design of the algorithms of the programs.

The current socio-political organization of the world's human societies is centred on state sovereignty and the rule of law. At the same time, economic lobbies and multinationals have tremendous power over politics and social organizations. In many cases, the governments are hostages of economic lobbies. These lobbies tend to be focused on short-term profit in opposition to long-term sustainability. Environmental laws and the efficiency of their enforcement are the result of continuous more or less transparent negotiations between groups of stakeholders, including government officials and economical lobbies, as well as political parties, the scientific community, the media and civil society. As a result of this often obscure negotiation process, environmental laws not only have severe gaps, but their enforcement is also occasionally unsatisfying. Dependent on regions and states, examples of gaps in environmental laws range from lack of regulation to control irregular uses of land and waters by individuals and enterprises, to uncomplete listing of hazardous substances leading to the improper disposal of harmful wastes. While governments, NGOs and scientists work on improving and updating the legislation, other individuals and enterprises, knowingly or unknowingly, take advantages of these gaps to continue practices degrading the environment and depleting natural resources.

Umbrella organizations such as the United Nations (UN) bodies and international NGOs have limited power as they rely on state sovereignty, and states are widely affected by various types and levels of influence by the economic lobbies. Different programs developed by international organizations attempt to skip the states and the problem of law enforcement by empowering local communities and individuals. Some of these projects can be found under the label "localization of the Sustainable Development Goals". They aim to empower local communities by sharing mainly educational resources (sometimes also financial and legal resources). Despite an encouraging start, the success of these initiatives remains limited, and is always restricted by dominant actors inside nation-states, be it the government, the military, the economic corporations or others.

In the recent years, social media have taken a tremendous importance in designing the global discourses. On the one hand, this can seem worrying, as it gives immense socio-political powers to social media platforms such as Facebook, WeChat and Line, which are enterprises seeking profits, and not social mediators of conflicts. Yet, by becoming the stage of conflicts and the main space where some groups share information and express their worries and discontentment, they are also called to play a much larger role than keeping the servers running. Any small change in the design of the algorithms can have a massive impact on who sees what kind of information, and so on shaping public opinion. On the other hand, social media platforms have also been used to give a voice to underrepresented people, and they are a central tool of activism, including environmental activism. They are not only used by individual users, but also by governments, NGOs, research institutes and UN bodies to actively try to influence the global discourse "for the better", for example for biodiversity conservation campaigns and for the promotion of nature-based solutions. A significant part of environmental NGOs funding, at least in Western countries, comes from individual members, by membership fees, donations or legs. Social media campaigning is a crucial tool for these NGOs to reach to potential new members and to raise awareness about the issue they are working on.

The landscape of nowadays' applied ethical thinking has widely changed from a hundred years ago. Individual ethical agents (taken to be any conscious human being) are connected and exchanging with more conversation partners than ever before. The consequences of individual agents' actions are also taking place at a very large scale. We are all connected to other human beings through our shared environment, through the economic ties of trade and through the traditional media (newspapers, radio, TV) and the Internet and social media. In other words, we are interacting at an unprecedented scale, and any interaction has social and often environmental consequences. Therefore, we need some common guidelines to orientate our behaviour beyond our immediate surroundings. We need some tools to make sense of the social and environmental situation we live in, that grasp the multiple layers of complexity without sacrificing sociocultural diversity for the sake of explanatory efficiency. We also need some shared moral standards and a consensual normative direction towards which we can direct common actions. And we need some tools to assess what is our individual role in this globally interconnected world. The purpose of this book is to draft a tentative motivational framework that contributes to address these needs.

1.2 Current dominant key ideas

We think, take decisions and justify our decisions with ideas and concepts, situated in a specific worldview. Because they are immaterial and complex to grasp, they may seem inconsequential and secondary to "real" and measurable facts and actions of the concrete material world. Yet, they are not only the hand of the compass we use to orientate ourselves when taking an ethical decision. They are also the design of the compass itself, the symbols on its face and, finally, they are the reasons that make us put out the compass and look at it, or leave it in our pocket and turn away. As I attempt to work with ideas, concepts and worldviews in an effort that would modify some existing ones towards other potentially "better" ones, I first need to describe briefly what are the dominant concepts and worldviews underlying the social and environmental situations we are in today.

1.2.1 Nature and resources

A key concept in the dominant worldview on the relationship that human beings have with nature is the idea of resources. The environment – an umbrella concept wide enough to cover multiple relations, flows and natural elements at diverse scales, from microorganisms to the climate – provides resources to human beings. These resources are there for us, and we need to exploit them, to use them and to manage them in the best way we can. It would be a pity if we waste them. Accordingly, the environment is seen as a passive chemico-physical receptacle replaceable by technology. If we come to the point where we do not have a specific resource anymore, we just need to use a technological substitute to it. There is no other value to the environment that for it to be used by us. Generally, this usage is understood as merely the consumption of material resources. Sometimes, it also includes the notion of non-material benefits such as aesthetical and recreational ones.

The concept of ecosystem services grew to be an essential part of the mainstream debates on the ways humans can and should use the natural resources smartly. They refer to the benefits that humans gain from "properly" functioning ecosystems. Ecosystem services have been divided into four categories popularized by the Millennium Ecosystem Assessment. According to it, drinking water is a provisioning service, climate cycles are regulating services, natural pollination of human-grown crops is a supporting service and a beautiful landscape is a cultural service for its spiritual and recreational benefits. Often, monetary values are assigned to each specific service in an attempt to help decision-makers making trade-offs and economically efficient policies. This commodification of ecosystem services allows them to be sold to the private sector and to be traded in markets.

This brings us to the question of assigning *what type of value* to *what entities* – a central axiological debate in environmental ethics. One of the main criticisms addressed to the concept of ecosystem services as used in the mainstream to foster commodification of natural elements is precisely that it reduces these elements to having only an instrumental value. According to the view that is largely denounced in environmental ethics, to attribute solely instrumental values to natural elements providing ecosystem services amounts to judging them to be exchangeable for something else (generally a technological substitute) that can provide equivalent services. For example, this would mean that if technology was able to compensate totally the "work" of insect pollinators, then the dramatic extinction of most of the pollinator species would no longer be a problem. This would at least hold for the ugly and unnoticeable species that enjoy no fame nor affection from human beings, and so that do not provide any "recreational" services.

This axiological debate evolved with the aim to be more inclusive of diverse values and cultural worldviews, and also of indigenous and local knowledge systems. Other concepts emerged to address the biases of "ecosystem services", such as the idea of "nature's contributions to people" (Pascual et al. 2017). In contemporary applied environmental philosophy, philosophers have been suggesting diverse ways to bypass these problems (Oruka 1996; Clowney and Mosto 2009; Diaconu and Kirloskar-Steinbach 2020). One way to clarify the debate is to distinguish three main categories in which nonhuman natural elements have value for human beings (Jax et al. 2013). They have first a fundamental value because they provide the basic conditions for life on earth. They also provide the basic conditions for a *good human life*, and that is their eudemonistic value. Once the fundamental and the eudemonistic values are clarified, the concept of instrumental value can be used exclusively for elements that are a means to something else, that is, elements that are replaceable. Another way in which philosophers addressed this issue, especially in the English-speaking literature, is to argue for an intrinsic or inherent moral value. A natural element that has an intrinsic value is valuable for *its own sake*. In other words, it is claimed to have value independently from its relation to humans and to human well-being. This has a strong deontological implication as intrinsic value entails a direct moral obligation towards natural elements holding it.

This short summary does not give enough credit to the complexity and variety of theories supporting these different types of values. Yet, two important remarks must be made here. First, these attempts to assigning values to nonhuman natural elements are often characterized by an implicit ubiquitous dichotomy between human beings and nature. This human-nature dualism is the target of criticisms by ecofeminist philosophers who argue that such a duality has been almost always coming along with a hierarchy. According to this criticized dualism, the natural world is associated with unpredictably dangerous and fascinating powers that might destroy humanity, if not controlled and tamed by the intelligent and cultivated human beings. This view is sometimes informed by monotheist religions, such as Judaism, Christianity and Islam, in which nature was given to humans for their own benefits. Environmental ethics theories that argue for an "earth stewardship" are also often rooted in such a clear demarcation between human beings and nature. In these views, there is an essential difference between human beings and the natural world. Because of their higher capacities and essential qualities, humans are *in charge of* the morally passive natural world surrounding them.

Second, most of the debate obscures the question of *who* is to assign these values. Are they to be found in the natural elements? Or, if they are assigned by human beings, who among human beings have the authority to do so and why? And in both cases, *how* are we to find or decide of them? A common reply to these questions is that values, especially intrinsic and inherent values, must be assigned to nonhuman natural beings by means of convincing rational arguments (Jax et al. 2013). This collides with the problem of pluralism of worldviews and with the problem of diversity in what is considered to be "rational".

1.2.2 Pluralism of worldviews and extremism

A high diversity of worldviews coexist in our interconnected world. A worldview is a comprehensive conception of the world from a specific standing point. Along with globalization and the increase of interactions between people, it becomes more obvious for all that not everybody holds a similar worldview. At the very least, it is clear that significant differences of opinion exist, even between people who appear to be knowledgeable and sincere. Honest and well-intentioned individuals who are similarly seeking for the truth on a specific matter can reach drastically different and incompatible conclusions. I briefly discussed the "dominant worldview" on the relationship between humans and nature as mediated by concepts like resources and ecosystems services. Countless other worldviews on the relation between humans and nature exist. One may think of the worldviews based on religious traditions, on indigenous local sets of beliefs – often characterized with attachment to the land – and on recent conceptions of the human-nature relation, such as Deep Ecology movements.

Each of these worldviews has concrete consequences on the world. Because they influence or even determinate decision-making of individuals, followed by actions that have consequences on the world. For example, the resource-based worldview is supporting an exploitation of natural elements regardless of their symbolic, religious or aesthetic values. On the contrary, a worldview based on Shintoism will isolate and protect natural areas deemed to be sacred even if they happen to be in the centre of a city where land prices reach peaks. Moreover, the same individual can hold worldviews that seem to be incompatible, without even noticing their seemingly contradictory aspects.

Some worldviews and their corresponding programs for action are compatible to a certain extent. However, many are conflicting on details, and some even face strong opposition on central propositions and directions of their lines of actions. Proponents of each worldview are constantly innovating their communication strategies to keep their members convinced and to persuade potentially interested members. Indeed, if compatibility between opposing worldviews is possible to a certain extent for some people, it remains limited. Members have a limited amount of financial resources and of time they can invest in one worldview. In most cases, a strongly convinced membership basis is more worthwhile for the advocates defending this worldview than a sparkled influence on a high diversity of sympathizers.

Leaders and spokespersons of worldviews are working in a highly competitive field. And the most powerful tool to successfully keep one's members captivated and to actively increase a group's membership is to play with emotions. Emotions of fear and hate are very useful to shatter people's most intimate beliefs and then to swiftly influence them with comforting speeches about positivity and love. These are no secret in the communication industry, and the refinement of these strategies reaches new heights when it comes to extremism (including the sadly famous multiple religious extremisms). Chaiwat Satha-Anand describes how romancing extremism is powerful: "The extreme is so magical that it blinds people attracted to it" (Satha-Anand 2018).

In philosophy of moral value pluralism, extremists are often referred to as exclusivists. An exclusivist refers to a person believing in only one basic worldview to be the truth or the closest to the truth than any other worldviews. If an exclusivist posture can be moral or not is strongly debated. Some philosophers argue that an exclusivist is under an obligation to engage in a belief assessment, because as their beliefs have consequences, they have a moral obligation to maximize truth (Forsberg 2007; Basinger 2020). And maximizing truth can be done only if each individual attempts to solve the conflict between incompatible contradictory beliefs.

The debate then shifts to question who shoulders the burden of proof of equal epistemic footing. Indeed, why would I question my own worldview if I judge that the proponents of the other conflicting view are not on equal epistemic footing as I am, because they are less educated, they do not have access to the information that I have, I judge them as having a lower capacity to think or they are essentially not equal to me because, for example, I am part of the "chosen ones" in my own worldview. This justification mirrors the "magic of the extreme" discussed by Chaiwat Satha-Anand. To believe that one has a special higher footing than others is complaisant and emotionally comfortable and satisfying. On top of this, when challenged by other positions, one often feels threatened and reacts by an emotional crystallization around some central beliefs that seem at the heart of one's identity. In many cultures, strength is equivalent to being strongly self-confident, and the latter might amount to holding harder foundational challenged beliefs without any apparent doubt.

Can epistemic peer conflict be resolved through dialogue and belief assessment? It may very well depend on the importance given to the specific challenged belief. Some beliefs are composing the very foundation of one's worldview, they are the givens in the specific belief system, by definition, and they can hardly be questioned, as the whole system built upon them would collapse in case of doubt. Jerome Gellmann refers to these beliefs as "rock-bottom beliefs" (Gellmann 1998). He argues that except these rock-bottom beliefs, all the other beliefs are subject to obligatory belief assessment when confronted to a diversity of other opinions. Even though, what criteria should be used to rule out certain options and comfort others? That is where concepts such as rationalism, self-consistency and comprehensiveness come into play. Even if within one specific community, all members share an implicit consensus about these tools, what if the rational dissection of the conflicting points regresses until rock-bottom beliefs? The disagreement might seem to be in a dead end that could have severe consequences on the world.

In sum, we can make the descriptive assertion that *there is* a high diversity of worldviews that may conflict with each other, and that even when "superficial" disagreements seem solvable with rational dialogue and self-consistency inquiry, *there are* still multiple ongoing conflicts rooted in rock-bottom beliefs. From here on, I thus take for granted these two facts, and the assumption that conflicting worldviews are not to be hierarchized. In view of this value and worldview pluralism, I adopt the precautionary posture that *one single truth* does not exist and that extreme cautions must be taken when judging diverse value-charged worldviews.

1.2.3 Ecological scientism and consensual rationalism

To deal with questions of environmental ethics at the global level, it is crucial to first clarify the relation between values and science. Indeed, environmental ethics lies at the complex intersection between science and ethics. In the global context of pluralism of worldviews, the distinction between scientific facts and situated value judgements is a key to prevent the imposition of one dominant worldview on others. Setting aside metaphysical questions regarding the foundations of reality, it is hardly contestable that there is a reality "out there", which can be referred to as "the world", "nature" or even "the environment". These umbrella terms attempt to cover the bundle of facts, phenomena and states of affairs that *are* and that surround us. Yet, there is an irreducible gap between this neutral objective reality and what we can perceive and access from our limited standpoint. To the best of our capacity and knowledge, we translate what we perceive into scientific truths using conventions that can be scientific, logic, linguistic or else. These so-called scientific facts are the theories that can best approximate what we can access of the reality according to the scientific consensus at a given time. They can be recorded in reports and journals, filtered by decision-making of the authors and editors. This filter makes the difference between what we *can* say and express (an epistemological question), and what we actually say and do (a conative question). Far from being neutral, actions and speech are strongly influenced by norms, values and meanings, and they are determined by the decision-making processes of individual phenomenological agents.

Two remarks must be made here. First, meanings require thinkers, and values require valuers. They are not facts existing outside of our thoughts, but only beliefs that depend on our experience. Most environmental philosophers recognize this point, even when they use the term of intrinsic value of nature. For example, J. Baird Callicott gives an account of anthropogenic intrinsic value of nature: "The *source* of all value is human consciousness, but it by no means follows that the *locus* of all values is consciousness itself", and "value is, to be sure, humanly conferred, but not necessarily homocentric" (Callicott 1986). Even authors that argue that "some values are objectively there" tend to accept that these "autonomous intrinsic values" of nature are "discovered" by valuers (Rolston 2012).

Second, perceivers, valuers, thinkers and believers are *always* individuals. Yet meanings and beliefs are built through interactions with other people. We understand things in the world and value them within the frame of a particular worldview that we learn and appropriate in the process of growing up and living together with others in a specific sociocultural context. Worldviews are the frame and lenses through which one perceives, makes sense of and acts on the world. They are set of meanings, values, beliefs, habits, patterns, etc. They are partially invisible to their beholders, but crucially influence their decision-making and ways of life. Therefore, it is hard to acknowledge, study and understand them. Nevertheless, noticing them is a must, because worldviews are *not* neutral scientific facts. Worldviews are held by individuals and groups. Worldviews are dominating in specific milieus, shaping how people live, think and act. The fact that some dominant worldviews tend to be shared by the members of a community makes it even harder to notice them, as they are not only one's individual frame, but a frame accepted by one's peers.

When discussing environmental problems in policy-making and mainstream media, naturalism and ecological scientism are the dominant paradigms. Naturalism is the worldview that only "natural" laws and forces govern the structure of the universe. In this context, "natural" is the contrary of spiritual. Ontological and metaphysical naturalisms argue that "all spatiotemporal entities must be identical to or metaphysically constituted by physical entities" (Papineau 2020). It excludes non-physical entities (e.g. spirits and gods) in the physical causal chains. Methodological naturalism is a methodological claim about how we acquire knowledge. It seeks explanations for events and scientific hypotheses with reference exclusively to natural causes and events.

Scientism is an ideology that takes naturalism one step further. It argues that societies should determine epistemological and *normative* values by science. In the discussions about environmental problems, ecological scientism refers to the idea that the environmental crisis is not ethical, but only a factual question coming from our partial ignorance. Species extinction and climate change are "accidents" that can be solved by understanding better the technical functioning of the Earth system (Gorke 2003). Proponents of ecological scientism adopt a posture of strong technical optimism, as they expect the progress of sciences and technology to bring us all the solutions needed. No reassessment and questioning of our current worldviews and ethical values are required. According to this view, my project here is hopeless as environmental problems are to be solved by engineering solutions and not by investigating normative premises. Yet, this position sweeps under the carpet the normative premises underlying the view of ecological scientism itself. Indeed, both ecological scientism and, to a certain extent, naturalism justify normative premises by judging them "rational".

Then, what is rational reasoning? Usually it is related to consistency, that is, the absence of contradiction between the elements of the reasoning. Yet, even if we accept this answer, absurd theories can be built on dubious premises while maintaining an internal consistency. Consistency is nevertheless a crucial element of "rationality", especially in the field of philosophy, because a philosopher who notices a contradiction internal to a system is expected to regard it as a serious objection. Another serious type of objection is when a consequence of an ethical system seems wrong, which rests on calls to "common sense" and on the untouchable set of principles that are deemed to be "rationally consensual" in a specific sociocultural group.

Actually, rationality also depends on a group sharing an internally consensual normative understanding of reason and logic, and in the field of ethical theory, of basic values and intuitive ideas about what is wrong. For example, people are rationally committed to ecological scientism if and only if they are similarly committed to its underlying normatively charged premises of ontological naturalism and technical optimism. Most philosophers are building theories of justice and ethics on an implicit basis of "some principle as a principle of rationality and then claiming that we are all committed to this principle" (Lehrer and Wagner 1981). They assume a rational consensus from the start. This is problematic because they fail to show that the individuals in a specific group (a society, a community or a readership) are actually consensually committed to these implicit premises.

This observation echoes the criticism by Helen Longino in her investigation of the underlying premises held by the scientific community: "Idiosyncratic values are suppressed, while values held by all members are invisible (as values, interests, or ideology)" (Longino 1995). Because they are invisible, they are also unavailable for discussion and debates. It follows that in a specific group, the values and worldviews of the dominant majority will unknowingly be imposed on the whole group as their own theory of rationality. Objections that use a different grammar and rest on an incompatible worldview will be discarded as irrelevant to the conversation and deemed "irrational".

What is needed practically is precisely a method to guide our actions when there is conflictual dissensus. Individuals sharing similar information about language, ethics and science are more likely to commit to their own consensual theory of rationality. Lehrer and Wagner refer to this form of consensus as a "rational amalgamation of the information individuals possess" (Lehrer and Wagner 1981). What if two different groups each reach such an internal consensus, but still conflict with each other? As Leherer and Wagner write: "The fact that both groups proceed in *formally* identical ways to aggregate their opinions and evaluations does not, in any way, show that their original opinions and evaluations were equally meritorious" (Lehrer and Wagner 1981). This brings us back to the former question of how to deal with individuals and groups identically strongly committed to the rock-bottom beliefs of different conflicting worldviews. It is important here to point out the fact that what is consensual and socially rational does not always coincide with what is moral. Considering these points, we can assume that consensual rationality criteria such as internal consistency and comprehensiveness can help solving some disagreements related to beliefs that are not foundational and rock-bottom. But they leave us empty-handed to address conflicts between opposing worldviews, values and theories of rationality rooted in rock-bottom beliefs without being partial or begging the question.

This type of deeply rooted disputes emerges a lot in the case of environmental problems, because they bring together people who live in different socio-political environments. This is the case at the local scale, when the installation of a pipeline crossing indigenous people's lands brings together business partners, native people, environmental and social activists, media and the police; or when drought-induced forest fires force people out of their houses regardless of their socioeconomic backgrounds. The diverse worldviews embodied in this multitude of individuals clash. This is also the case at the international level, when diplomats and NGOs meet to discuss an agreement of international environmental law and appear to be talking past each other.

Up to now, scientific rationale has been providing the main reliable consensual basis for negotiation. The Intergovernmental Panel on Climate Change (IPCC) and the Intergovernmental Science-Policy Platform on Biodiversity and Ecosystem Services (IPBES) are two examples of research networking platforms established by the UN to provide an "objective, scientific view" on the state of knowledge on climate change and biodiversity and their political and economic impacts. Besides writing scientific reports, they also compose summaries for policy-makers that are reviewed line by line by delegates from different countries and scientific bodies, and that are presenting a scientific politically consensual basis for further negotiations. These apparently consensual tools are key in the success of negotiations, not only at the international level, but also at the regional level, where they are often used as reference points in the establishment of national, regional and local policies. Thanks to these international research networking platforms, decision-makers are quite well equipped when it comes to scientific assessments and guidelines for action. More and more, international research groups are also trying to be inclusive and to collaborate with indigenous knowledge groups, enlarging the range of worldviews their reports can be compatible with.

Such international organisms have understood that in reality, what is needed to tackle most of current environmental problems is not a complete consensual agreement on the rock-bottom beliefs and worldviews. Such an "agreement" would not be realistic nor desirable. What is needed is sets of tools that are just consensual enough to support dialogue and common action. The success of the concept of human rights is an example of how a concept, vague enough to be interpreted in very diverse ways in different worldviews and so reach a "global" consensus, can have important ethical consequences. As it has been thoughtfully discussed during the last century, the concept of human rights itself means very different things and carries highly diverse nuances in different socio-political contexts and linguistic universes. The interpretation of the human rights covers a surprisingly wide range across the globe. Nevertheless, they are agreed upon and provide a common positive basis for collaboration and actions. They impact strongly the landscape of theories and reasoning and are moving the burden of justification from the "victims" to the violators of the conventional human rights. They also paved the way for a new moral identity as "human being" on which this book leans. With these practical considerations in mind, this book proposes to draft a motivational framework inclusive for diverse ideas of the environment and nature.

1.3 Pragmatism and need for consensus

It must be clear by now that I take a pragmatic stance in my reasoning, as I discuss the normative aspects of real-world environmental problems in the context of contemporary social realities. Yet, I want to clarify here what it entails and how it unfolds.

1.3.1 *Pragmatism, moral minimalism, and rejection of one universal environmental ethics*

Pragmatism is characterized by three mutually reinforcing ideas (Dieleman, Rondel, and Voparil 2017). First, it prioritizes real-world problems such as environmental problems ahead of abstract principles. Second, it holds a distrust of a priori theorizing, because it often simply imposes a dominant consensual state of affairs without questioning its underlying premises, as we saw earlier. Third, it recognizes pluralism, both in respect to what is ethics and what is right, and in respect to how to solve real-world ethical problems. I also previously recognized the pluralism of value as valuable in itself.

After investigating the question of justice pragmatically, Nancy Fraser finds three abnormalities that she describes with the three questions of What, Who and How (Fraser 2008). To start with, she notices that there is no shared view on "what" justice is. We have already observed that aspect when discussing the pluralism of worldviews encompassing, opposing and conflicting ideas of what ethics is. Then, she continues by saying that the "who" of justice is not clear, nor consensual either: Who deserves consideration and who is entitled to equal concern? This question is especially flagrant when it comes to environmental problems. They go beyond disputes internal to territorial states. Depending on the position of the party, they frame the question as a matter of domestic territorial "who" – jobs for local people in the coal mines, urban air pollution, etc. – or as regional, transnational or even global "who" – climate change, future generations, etc. The whole discussion around the exportation of environmental and social externalities questions the "who" of ethics. Finally, Fraser points out that there is no shared idea on the "how". What procedures and according to what guidelines should we solve the two previous questions? For Fraser, these three nods of abnormalities and the ambiguity surrounding them are the underlying sources of many disputes. These three nods can be found in my book; the "what" is addressed by the question of sustainability, the "who" by the idea of responsibility and the "how" by tentative pragmatic guidelines.

The difficulties surrounding the "what" of justice and ethics were already exposed in the discussion of the pluralism of worldviews and the limits of rationalism. The "who" of justice is questioning the membership rule of a group deserving equal consideration. Historically, answers to this question have used diverse criteria such as social class (e.g. aristocracy), richness or education (e.g. censitary suffrage), race (e.g. apartheid) and gender (e.g. male suffrage). Recently, exclusionary nationalism argues for the nation to be the limit of the range of justice. All these criteria seem hardly acceptable in the context of global environmental problems affecting primarily the poorest populations in the world. Philosophers also appealed to a principle of humanism rooted on a criterion of personhood. Disparate individuals become subjects of justice by virtue of their common possession of humane features. Descriptions of humane features vary but tend to converge around characteristics such as the capacity to think, to take decisions, to pursue an idea of good and to suffer from moral injury. The main objection to this view is that it "accords standing indiscriminately to everyone in respect to everything" (Fraser 2008). Then, in certain cases, voices of the victims are muffled under the dominant voices of the oppressive majority.

Closer to debates in environmental ethics, other philosophers argue for the all-affected principle based on transnational social relations of interdependence. If people are connected through causal relationships, then they would be equally entitled to ethical considerations. Yet, as global environmental problems such as greenhouse gases emissions leading to climate change show, the whole planet is interconnected. Thus, the criterion does not resist its own basis of interdependence, as we are all to a certain extent affected and affecting everybody else through a globalist domino effect.

Fraser proposes an "all-subjected principle" according to which "all those who are subject to a given governance structure have moral standing as subjects of justice in relation to it" (2008, 52–53). She understands subjection broadly, including any person "subject to the coercive power of non–state and transstate forms of governmentality". As we can suspect, determining the "who" is closely intertwined with problematizing the "how". Fraser rejects both the drafting of principles of ethics and justice done by the elite, and by sciences. She advocates a dialogical process between members of the civil society and a "dynamic, interactive" formal institutional framework. This would imply not only valorizing reflexive ethical reasoning by most members, but also the development of a shared common sense. Recent pragmatic positions often have in common a discussion about the revalorization of the diversity of experiences – including the experiences of the oppressed and the victims – and of the cultivation of self-criticism and affective sentiments of responsiveness and empowerment. What is needed is not a theory and abstract normative principles, but it is an "orientation, an affective sensibility, and set of moral and methodological priorities" for responding to concrete ethical problems (Voparil 2018).

A consequence of my pragmatic posture is to adopt a certain moral minimalism, which is required for environmental ethics to have a global reach. Detailed moral intuitions vary greatly depending on cultures and religious backgrounds. Thus, a system of ethics reaching up to intricate circumstantial questions that depend directly on fashion, norms and traditions would inevitably be situated in a specific spatiotemporal and sociocultural realm. In contrast, I aim to contribute to the construction of a global framework built on premises that can be widely consensual to be accepted by individuals and groups with diverse backgrounds and allow us to develop a discussion leading towards common actions. Such a motivational framework for environmental action needs to rest on premises that are part of the common sense of different cultures, even if they take various shades in each specific context. General principles are ambiguous, debatable, and will never unanimously convince the totality of the members of our large global community of ethical agents.

Nevertheless, we still need "some general principles, which could be employed by moral and political thinkers in order to systematize our central intuitions and serve as prima facie guidelines for action" (Virvidakis 2014). While some philosophers debate some details in complex moral theories, the environmental crisis and the subsequent humanitarian crisis are pressing us to take concrete action. In the end, "we are stuck with the necessity of passing moral judgement at the global level, and must create the motivational structure needed to sustain such judgements" (Risse 2012).

Minimalism recognizes that in practice, the judgement that an action is right or wrong by the agent or by an acquaintance of the agent will influence her[2] in doing or refraining from doing the action in question, which pragmatically justifies the building of a motivational ethical framework. It also entails the dismissal of foundational claims, and does not require the development of an ontology. Methodologically, this has implications on the justification of our ethical judgements. How can we justify our judgements without grounding them in an infallible basis? We can do so by appealing to coherence and comprehensiveness: by using a coherentist model of justification. In a coherentist model of justification, a belief or a set of beliefs is justifiably held if and only if the belief coheres with others and forms a coherent system. This coherent system does not need to be grounded on a truth – be it being, nothingness, deities, etc. – to be valuable. It is sufficient to be a sort of web of interrelated beliefs floating onto reality. In a coherent and minimalist project, neither deontological nor consequentialist elements can be neglected, even it might be more complicated to reach and develop in consensual tools from the former than the latter.

In view of the myriad of uncertainties that we face when we must take environmental decisions, one could easily argue that we do not know enough to take the most appropriate decision from a consequentialist perspective. Still, we have to make decisions and take actions every day for our own subsistence. Suspending our judgement will not help us in the context of the global environmental crisis, for we are in a situation of urgency. For instance, scientists generally agree that the implementation of actions to combat land degradation, climate change and biodiversity loss will become more difficult and costly over time (e.g. IPBES 2018). Immediate actions guided by available scientific evidence and the precautionary principle are highly desirable. Yet, we are not protected from making mistakes that will have detrimental ethical consequences. Nevertheless, in Martha Nussbaum's words, we must not "indulge in moral narcissism when we flagellate ourselves for our own errors", while others are suffering or will suffer from the consequences of our non-action (Nussbaum 1996). That is why I also take for granted a proactive forward-looking optimism. I assume that it is possible and normatively required to act for the best, given the present knowledge and circumstances even if, as always, there is a risk that the action has worse consequences than expected.

My pragmatic posture regarding global environmental ethics has another implication: the rejection of the possibility of *one* universal environmental ethics. Because of the recognition of the value of pluralism of worldviews and because of pragmatic considerations, the construction of a unique system of ethics that would be applicable universally is rejected as an imperialistic and problematic endeavour. Instead, I propose toolkits (the "how") to think a motivational framework with sustainability as a tentative direction for common actions (the "what"), and an account of responsibility for environmental harms (the "who").

1.3.2 Premises: we value human existence

Pragmatism led us to adopt moral minimalism, and thus to look for the "lowest common denominator" for a global ethics. I propose two minimalist premises that hopefully can be consensual enough to be agreed upon by individuals from many different cultures and backgrounds, if not by everybody. The first premise is "We value human existence". "We" refers to anybody engaging in the dialogue. As such, "we" is already validating the premise, because any potential reader or listener of this premise would be alive. By being alive, specifically by not having committed suicide, anybody must have something that they judge valuable in their own existence as individual, or at least in their existence as part of a network of human beings. Some will say that we bump into the difficulty of defining what "human existence" is. However, the purpose of this premise being to reach consensus, its readers will inevitably be human beings and so must have at least a non-problematized understanding of what human existence is. Such an intuitive understanding is enough for the sake of my argumentation.

Then, one could ask what is valued in human existence. This opens an abyss of questions and possible replies that are generally deeply rooted in the specific worldview and rock-bottom beliefs of the individuals speaking. Giving an answer to this question would make us quickly drift away from minimalism into an ontological enterprise, which is not desirable. Without entering into details on what human life is or should be, we can be satisfied with a consensual agreement on the fact that human survival is positive and good, and that the possibility for humane existence should be maintained.

This first premise is followed by a second one that is almost a corollary, namely: "A healthy environment and a meaningful milieu are necessary conditions for human existence." Healthy means, first and foremost, healthy for human beings. I do not mean here to personify the environment as a being that can be healthy or sick. What is healthy for human beings is also very ambiguous and depends on what is valued in human existence, on the physical and mental vulnerabilities of the individual human in question and on the worldviews endorsed by her and her community. I will describe in more details what I mean by the "environment" and by the "milieu" in Chapter 2.

A general line of argumentation can follow from these two premises: If a healthy environment and a meaningful milieu are necessary conditions for human existence, then environmental sustainability is a necessary condition for human survival. Environmental sustainability can be understood in its most common form, namely that humans use ecosystems at a pace and in a way allowing their self-regulation. In Chapter 3, I will redefine sustainability in more details following the implications of the concept of the milieu developed in Chapter 2.

Methodological implications can also be derived from these two minimalist premises. In short, environmental sustainability is a requirement for human existence, as it maintains a healthy environment and a meaningful milieu. Then, social sustainability, namely social stability and the near absence of large-scale conflicts, is often recognized as a necessary condition for environmental sustainability. For instance, in cases of war, conflicts and other socio-political threats, the national budget dedicated to environmental protection is often one of the first to be cut down, as political security is given priority (Jensen and Halle 2009). Likewise, imposing environmental laws on disagreeing subjects is likely to lead to social unrest as diverse ranges of protests might follow. Then, we can suppose that if individuals are morally convinced of and committed by themselves – that is, non-coercively – to environmental sustainability, then social sustainability is more probable to be achieved, and vice versa.

How do individuals become convinced that a course of action is better than another? Every individual is embodied, situated in time, space and culture, and becomes convinced by balancing ethical values. Correspondingly to the coherentist model of justification, balancing values and principles to reach a temporary reflective equilibrium (Rawls 2005; Ron 2006; Cath 2016) seems to be a common enough method of taking individual decisions. Ethical worldviews and legal systems carried by cultures influence the individuals in balancing their values. Therefore, to reach environmental sustainability, we need to understand how individuals become convinced to take an ethical action, and the ethical worldviews influencing them. Ultimately, the goal of the motivational framework for environmental ethics that I attempt to contribute building is to have an influence on these ethical worldviews.

To build conceptual tools and justifications that can efficiently convince individuals and groups to adopt environmentally friendly ways of life, we need to understand the individual decision-making processes and the worldviews informing them in the case of environmental ethical actions. This can be done by dialogical philosophical reasoning. Plus, as my project is pragmatic, it is crucial to keep listening to the voices of individuals working in this domain outside of the academic world. The philosophical reasoning in this book is importantly informed by experiences and discussions with the environmental advocate community buzzing beyond the dusty libraries and abstract philosophical debates. Therefore, it is dialogical, because it is developed in conversation with other thinkers from different social status, disciplines, and linguistic, cultural and economic backgrounds.

1.3.3 Obstacles to environmental ethical action

To understand better how individuals get convinced of taking actions supporting environmental sustainability, we need to clarify the obstacles to environmental ethics action. Most of the time, individuals who have at least basic knowledge about environmental sustainability also agree on its importance. But somehow, something is preventing them from acting on their positive intuitions. Through self-examination and in-depth discussions, I identified diverse obstacles diverting individuals from taking environmentally friendly actions. Grasping what these obstacles are helps designing solutions to overcome them. They furnish more specific points to be tackled by a motivational framework for environmental ethics.

When an individual or a group wants to take an environmental ethical action, they may face psychological obstacles and social obstacles. Psychological obstacles to environmental ethical action include epistemological ones such as the absence of concepts, a limited blindness that may be linked to a dogma and a lack of conceptual tools and justifications. Without some key concepts and words, we are limited in how we think and how we express opinions. We fail to find the words to describe, explain and justify emotional moral intuitions. For example, the nuances of the translations of the concept of "sustainability" in different languages impact on how it is normatively used and shape the discourses around it. In English, it echoes the idea of support over an unlimited time frame. In contrast, in French, the equivalent word "durabilité" does not have any nuance of support and refers to a defined time span over which something lasts. Such slight differences have domino effect on how the principles and policies designed around these key concepts are understood by the public and the policy-makers. Further, we are sometimes unable to question some aspects of our ways of life because we believe in some rock-bottom beliefs that limit our view of the world. This blindness can be a choice and appear dogmatic from the outside. These blind spots are problematic when some environmentally detrimental actions are regarded as essential and unquestionable by a community or an individual. Even when words exist, we might fail to articulate ideas and moral intuitions into a convincing argumentation, and we might lose face during debates with opponents. Combined with social pressure, this obstacle can prevent otherwise motivated individuals from taking action.

Psychological obstacles also include emotional aspects such as lethargy, apathy and lack of self-confidence. These emotional hindrances are often combined with socio-political obstacles such as social isolation, lack of resources, opposing laws, institutions and states, non-complying agents and finally threats to one's integrity and life (Jasper 1997). The lack of resources, including financial, educational and skills-related, affects the success and failure of pro-environmental projects. The socio-political context can also present obstacles to environmental ethical actions, for example through opposing laws, institutions and states, that are out of the reach of the individual. Finally, the agents can receive direct threats to their life or their relatives', especially in the field of environmental activism. These real-life obstacles drive agents to seriously balance risks and commitments and set the stage for the reasoning of my book to unfold.

1.4 Plan

My project is situated in the field of environmental political philosophy and ethics. I take a pragmatist perspective to think about ethics related to real-world environmental problems in the globalized context of pluralism of worldviews. My overarching objective is to build a motivational framework that supports sustainable behaviours and that can be widely consensual globally by being compatible with different worldviews. As a starting point, I take the idea of a dynamic and adaptable relation between human beings and their environment mediated by the milieu. I am inspired by Watsuji Tetsurō's idea of non-atomistic individual continuously in cyclic relations with her milieu.

The book is divided into five chapters, including the introduction in which I describe the background, objectives and methodology. In Chapter 1, I argued that we are all facing the global environmental problems together, and so that we need to understand each other and our relation to the world to engage in common solutions. The main research question that enfolds the whole book is: Within an interconnected world and in our globalized context of pluralism of worldviews, on what grounds can we develop consensus to take common pro-environmental actions and drive transformative changes? The thesis I defend in answer to this question is that the idea of milieu can ground consensual ethics of sustainability. Specifically, the milieu is a useful tool to think our relationship with the environment, to root a conception of sustainability towards which direct our common actions and to anchor an account of individual responsibility for environmental harm.

In Chapter 2, I explore the question of how do we see and shape the world. To do so, I develop the concept of milieu further and incorporate it within my three-level model. I present Watsuji's concept of milieu, which is the "environment" as it appears covered by web of significations and symbols from the standpoint of a subjective human. I show how, on the one hand, the medial matrix shapes individual human beings, while, on the other hand, individuals shape the milieu by their imprints. The research question framing this chapter is: How do we develop worldviews and practices to use our surroundings (our milieu) in particular ways? The thesis that I defend in answer to this question is that we are shaped by the milieu through participatory sense-making, borrowing meanings to the cultural imaginary and constrained by practices (matrix); and we are shaping the milieu through the traces of our actions (imprint).

Chapter 3 addresses the questions of what do we want to sustain, in other words, what is sustainability as the desirable direction for our actions and practices. The research question structuring this chapter is: What do we want to sustain/maintain, and what are the normative implications of this understanding of sustainability in relation to the idea of milieu? To answer it, I develop a working definition of sustainability as the maintenance of the conditions of possibility of continuation of (1) self-determining flourishing human existences. It entails (2) maintaining the general processes of the global environment healthy to limit the possible harmful consequences of the conflicts of distribution and domination, and (3) cultivating meaningful, diverse and adaptable nurturing milieus. Then, I draw some normative implications from this definition, and I address some important objections, limits and priorities.

In Chapter 4, I explore what is our individual responsibility related to environmental problems and sustainability, and what it entails for the individual agents. The research question underlying this chapter is: What is individual responsibility for environmental harm? As an answer, I develop an account of individual responsibility for environmental harm, which is the result of balancing our contributory responsibility (imprint – wide-encompassing, including omissions and ways of life) and capacity responsibility (matrix – vulnerabilities, threshold of basic needs and structural political responsibility). I discuss individual responsibility for environmental harm through the lenses of the concept of milieu. Imprints are linked to the contributory individual responsibility, while the matrix constrains the readiness to be responsible. Then, I discuss what kind of reparative actions this account of responsibility encourages, and how we experience taking responsibility. In our interconnected world, many reparative actions require collaboration between individuals and between milieus, so limits and tentative pragmatic safeguards are introduced to prevent counterproductive interventionism.

Finally, in Chapter 5, I bring all these different insights together to answer the question that encompasses the whole book, namely on what grounds can we develop consensus to take common pro-environmental actions and drive transformative changes in the globalized context of pluralism of worldviews. I show that the idea of milieu, along with the account of individual responsibility in terms of milieu, can provide common grounds for a consensual ethics of sustainability.

The following table synthesizes the questions discussed in each chapter, and the thesis defended in answer them. Naturally, at this point of the introduction, the reader is not expected to understand each thesis of this table – otherwise, I could end the discussion here. Instead, I propose to use it as a roadmap to navigate through the book.

Table 1.1 Plan of the book

Chapter 1: Introduction – Within an interconnected world and a globalized context of pluralism of worldviews, on what grounds can we develop consensus to take common pro-environmental actions and drive transformative changes?

• Thesis: The idea of milieu can provide common grounds for a consensual ethics of sustainability.

Chapter 2: Milieu – How do we develop worldviews and practices to use our natural surroundings (our milieu) in particular ways?

• Thesis: We are shaped by the milieu through participatory sense-making, borrowing meanings to the cultural imaginary and constrained by practices (matrix); and we are shaping the milieu through the traces of our actions (imprint).

Chapter 3: Sustainability – What do we want to sustain/maintain, and what are the normative implications of this understanding of sustainability in relation to the idea of milieu?

• Thesis: Sustainability is the maintenance of the conditions of possibility of continuation of (1) self-determining flourishing human existences. It entails (2) maintaining the general processes of the global environment autonomous and healthy to limit the possible harmful consequences of the conflicts of distribution and domination, and (3) cultivating meaningful, diverse and adaptable nurturing milieus.

Chapter 4: Responsibility – What is individual responsibility for environmental harm?

• Thesis: Individual responsibility for environmental harm is the result of balancing our contributory responsibility (imprint – wide-encompassing, including omissions and lifestyles) and capacity responsibility (matrix – threshold of basic needs, and structural political responsibility).

Chapter 5: Conclusion

Notes

1 This book originated from a doctoral thesis defended in February 2020 at Kyoto University (Droz 2020).
2 The feminine is used as gender neutral.

Bibliography

Basinger, David. 2020, Winter. 'Religious Diversity (Pluralism)'. In *The Stanford Encyclopedia of Philosophy*, edited by Edward N. Zalta. Metaphysics Research Lab, Stanford University. https://plato.stanford.edu/archives/win2020/entries/religious-pluralism/.

Callicott, J. Baird. 1986. 'On the Intrinsic Value of Nonhuman Species'. In *The Preservation of Species*, edited by Bryan G. Norton. Princeton, NJ: Princeton University Press, 138–72.

Cath, Yuri. 2016, 19 May. 'Reflective Equilibrium'. In *The Oxford Handbook of Philosophical Methodology*. https://doi.org/10.1093/oxfordhb/9780199668779.013.32.

Clowney, David, and Patricia Mosto. 2009. *Earthcare: An Anthology in Environmental Ethics*. Lanham, MD: Rowman & Littlefield.

David Jensen and Silja Halle. 2009. *Protecting the Environment During Armed Conflict, An Inventory and Analysis of International Law, United Nations Environment Programme*. Nairobi, Kenya, ISBN: 978-92-807-3042-5. https://postconflict.unep.ch/publications/int_law.pdf

Diaconu, Madalina, and Monika Kirloskar-Steinbach. 2020. *Environmental Ethics: Cross-cultural Explorations*. Freiburg/München: Verlag Karl Alber.

Dieleman, Susan, David Rondel, and Christopher J. Voparil. 2017. *Pragmatism and Justice*. New York, USA: Oxford University Press.

Droz, Laÿna. 2020. *The Milieu as Common Grounds for Global Environmental Ethics*. PhD dissertation, Kyoto: Kyoto University.

Forsberg, Ellen-Marie. 2007. 'Value Pluralism and Coherentist Justification of Ethical Advice'. *Journal of Agricultural and Environmental Ethics* 20 (1): 81–97. https://doi.org/10.1007/s10806-006-9017-6.

Fraser, Nancy. 2008. 'Abnormal Justice'. *Critical Inquiry* 34 (3): 393–422. https://doi.org/10.1086/589478.

Gellmann, Jerome I. 1998. 'Epistemic Peer Conflict and Religious Belief: A Reply to Basinger'. *Faith and Philosophy* 15 (2): 229–35. https://doi.org/faithphil199815218.

Gorke, Martin. 2003. *The Death of Our Planet's Species: A Challenge to Ecology and Ethics*. Washington DC: Island Press.

IPBES. 2018. The IPBES Assessment Report on Land Degradation and Restoration. In *Secretariat of the Intergovernmental Science-Policy Platform on Biodiversity and Ecosystem Services*, edited by Montanarella, Luca, Scholes, Robert, and Brainich, Anastasia Bonn, Germany. 744 pages. https://doi.org/10.5281/zenodo.3237392.

———. 2019. The Global Assessment Report on Biodiversity and Ecosystems Services. In *Intergovernmental Science-Policy Platform on Biodiversity and Ecosystem Services*, edited by Eduardo S. Brondizio, Josef Settele, Sandra Díaz, and Hien Thu Ngo. Bonn: IPBES Secretariat, UNEP.

———. 2020. 'IPBES Workshop on Biodiversity and Pandemics'. In *Intergovernmental Platform on Biodiversity and Ecosystem Services*, edited by Peter Daszak, John Amuasi, Peter Buss, Carlos Das Neves, Heliana Dundarova, Yasha Feferholtz, Gabor Foldvari, David Hayman, Etinosa Igbinosa, Sandra Junglen, Thijs Kuiken, Qiyong Liu, Benjamin Roche, Gerardo Suzan, Marcela Uhart, Chadia Wannous, Katie Woolaston, Carlos Zambrana Torrelio, Nichole Barger, David Cooper, Tom De Meulenaer, Hans-Otto Poertner, Cristina Romanelli, Karen O'Brien, Paola Mosig Reidl, Unai Pascual, Peter Stoett, Hien Thu Ngo. Bonn: IPBES secretariat, doi:10.5281/zenodo.4147317.

Jasper, James M. 1997. *The Art of Moral Protest: Culture, Biography, and Creativity in Social Movements*. Chicago: University of Chicago Press.

Jax, Kurt, David N. Barton, Kai M. A. Chan, Rudolf de Groot, Ulrike Doyle, Uta Eser, Christoph Görg, et al. 2013. 'Ecosystem Services and Ethics'. *Ecological Economics* 93 (September): 260–68. https://doi.org/10.1016/j.ecolecon.2013.06.008.

Lehrer, Keith, and Carl Wagner. 1981. *Rational Consensus in Science and Society*. Boston, MA: D. Reidel.

Longino, Helen E. 1995. 'Gender, Politics, and the Theoretical Virtues'. *Synthese* 104 (3): 383–97.

Nussbaum, Martha. 1996. 'Double Moral Standards? A Response to Yael Tamir's "Hands Off Clitoridectomy," from the October/November 1996 Issue of

Boston Review.' Boston Critic, Inc. MIT Press, https://bostonreview.net/archives/BR21.5/nussbaum.html

Oruka, Odera. 1996. *Philosophy, Humanity and Ecology: Philosophy of Nature and Environmental Ethics*. DIANE Publishing.

Papineau, David. 2020, Summer. 'Naturalism'. In *The Stanford Encyclopedia of Philosophy*, edited by Edward N. Zalta. Metaphysics Research Lab, Stanford University. https://plato.stanford.edu/archives/sum2020/entries/naturalism/.

Pascual, Unai, Patricia Balvanera, Sandra Díaz, György Pataki, Eva Roth, Marie Stenseke, Robert T. Watson, et al. 2017. 'Valuing Nature's Contributions to People: The IPBES Approach'. *Current Opinion in Environmental Sustainability*, Open issue, part II, 26–27 (June): 7–16. https://doi.org/10.1016/j.cosust.2016.12.006.

Rawls, John. 2005. *A Theory of Justice: Original Edition*. Cambridge, MA: Harvard University Press.

Risse, Mathias. 2012. *Global Political Philosophy*. Palgrave Philosophy Today. Palgrave Macmillan UK. https://doi.org/10.1057/9781137283443.

Rolston, Holmes. 2012. *Environmental Ethics: Duties to and Values in the Natural World*. Philadelphia: Temple University Press.

Ron, Amit. 2006. 'Rawls as a Critical Theorist: Reflective Equilibrium after the "Deliberative Turn"'. *Philosophy & Social Criticism* 32 (2): 173–91. https://doi.org/10.1177/0191453706061091.

Satha-Anand, Chaiwat. 2018. 'Romancing Extremism?: Understanding the Violent Turn in Struggles for Rights, Responsibility and Justice'. *Plenary Presentation at the World Congress of Philosophy*, August 17th 2018, Beijing.

Virvidakis, Stelios. 2014. 'Moral Minimalism in the Political Realm'. In *Liberamicorum*, edited by Dutant Julien, Davide Fassio, and Anne Meylan. Geneva: University of Geneva, 860–77. https://www.unige.ch/lettres/philo/publications/engel/liberamicorum/virvidakis.pdf

Voparil, Christopher J. 2018. 'Social Justice Pragmatism: Toward a Working Program'. *Presentation at the World Congress of Philosophy in Beijing*, August 2018. Draft Paper given by the Author.

2 Milieu

I develop here a conceptual toolbox around the key concept of the milieu.[1] I start by situating it as the intermediary between the individual and the environment in my three-level model. Then, I review the literature around the concept of milieu and introduce the distinction between the milieu seen as a collective imprint and the milieu seen as an historical matrix. I explore further how the medial matrix shapes individual humans through dynamic and participatory processes of sense-making involving the cultural imaginary and practices. Finally, I investigate ethical action as a medial imprint.

2.1 Three-level model

I present here the three-level model of the environment, milieu and the individual. Then, I introduce the useful distinction between the perspective of the observer and the perspective of the phenomenological agent. Finally, I explore briefly the social construction of the self, as it is an essential relata to the idea of milieu.

2.1.1 Three levels: environment, milieu and the individual

The mainstream dominant worldview about the relation between humans and nature reduces it to these two main elements, placing them in a dualist opposition. The relation between humans and nature is then characterized in many different ways, based on hierarchy, domination, care, ownership, stewardship, etc. Sometimes, especially in Western classical dualism, the human-nature dualism conveniently overlaps the mind-body dualism. In other words, humans control and own nature in the same way the human mind is supposed to control and own the body.

This dualism has been widely criticized. I shortly give two examples of criticisms that converge despite coming from very different horizons. First, many East Asian traditions of thoughts are characterized by holism, embodiment and non-duality (Deguchi et al. 2021). According to this trend in East Asian philosophy, there is no mind-body or human-nature dualism. Humans are always embodied, and they should attempt to make "one" with the world.

Second, far more recently, ecofeminist philosophers have fervently criticized this dualism, seeing it as the common source of the relations of domination and abuse of humans over nature, and of men over women. According to pioneers of ecofeminism (Plumwood 1991), the worldview equating nature to emotions, bodies and womanhood is using the same reasoning pattern to justify their inferiority and the necessity to be controlled, tamed, managed and protected by the reason and minds of human males. A lot can be said about the limitations and worrying implications of such criticism. For instance, such a criticism risks reinforcing the naturalization and essentialization of what womanhood and manhood consist of and of their dichotomy (Gaard 2011). Plus, it seems to support an underlying assumption of binary genders, which might be widely consensual in some sociocultural contexts (MacCormack and Strathern 1980), but not in all, and which is now under scrutiny (Linstead and Brewis 2004). Again, such a short description does not give justice to the complex depths and multiplicity of these two families of criticisms, but it is enough to illustrate some of the limitations of the human–nature dualism.

Instead of this dual worldview opposing humans and nature, I develop a model based on three densely interconnected levels: the environment, the milieu and the human individual. These three levels can be imagined as concentric circles expanding around the human individual, or as the field of perception from the perspective of the individual. The reason why I anchor the model on the individual human is because it is the place of ethical agency. The individual is seeing the world, understanding it, making decisions about it and acting upon it always from her specific alive standing point. I believe it is impossible to entirely abstract the human individual from her situated, embodied and temporally limited existence. Moreover, such an abstraction would in no way be helpful in developing a pragmatic environmental ethics designed to support environmentally friendly decision-making and actions of living human individuals. Living human individuals are my only audience and conversation partners. They are also the only alive agents that can currently act and change their behaviours to improve the impacts that humanity as a whole has on the planet. Every time I use the terms "we", "individual" and "group", I thus refer to alive human beings who could be able to follow my reasoning, to human individuals and to groups of human beings.

Individual humans are always existing spatially and temporally in a specific environment that is, at least up to now for the overwhelming majority, situated somewhere on the surface of the Planet Earth. This spatiotemporal environment is the fundamental ground and receptacle for our existence as it provides us the necessary conditions for our biological life, such as food, air, water and atmospheric pressure. It is also the ground from which we think, imagine, build meanings and borrow shapes and relations to make sense of our existence. We are all living in this spatiotemporal environment, along with other human beings, plants, animals and an impressive range of diverse living organisms.

Figure 2.1 shows the three-level model. At the base of the circular pyramid lies the spatiotemporal environment. For the sake of visibility and to clarify the perspective of the moral agent, I isolate the two other levels of the milieu and the individual, but actually, the higher levels are simply part of the lower ones. In other words, the three levels can collapse into a single one, namely, the world. These three levels are not hierarchized. Instead, they correspond to different scales of zooming on the interactions between an individual human and her world, the latter including the spatiotemporal environment, always mediated by the milieu, as we will see.

For the sake of our discussion, the spatiotemporal environment can be delimited to the Planet Earth and all the elements directly impacting it (the solar system, etc.). It includes other living organisms and non-living things. It is expanded spatially far beyond what is accessible to one single human individual and encompasses at least the entirety of the humanely accessible and known surface of our planet. It also expands temporally as it existed before our individual births, and will remain existing after our individual deaths. We can also suppose that the environment existed before human beings and will exist after our disappearance as a species, but it is not necessary to settle or approach this question for our discussion.

When thinking about environmental ethics, we need to situate the ethical agent *within* the world, imbricated in self-other relations, and observer of other-other relations. Because we are moral agents, any relation and interaction

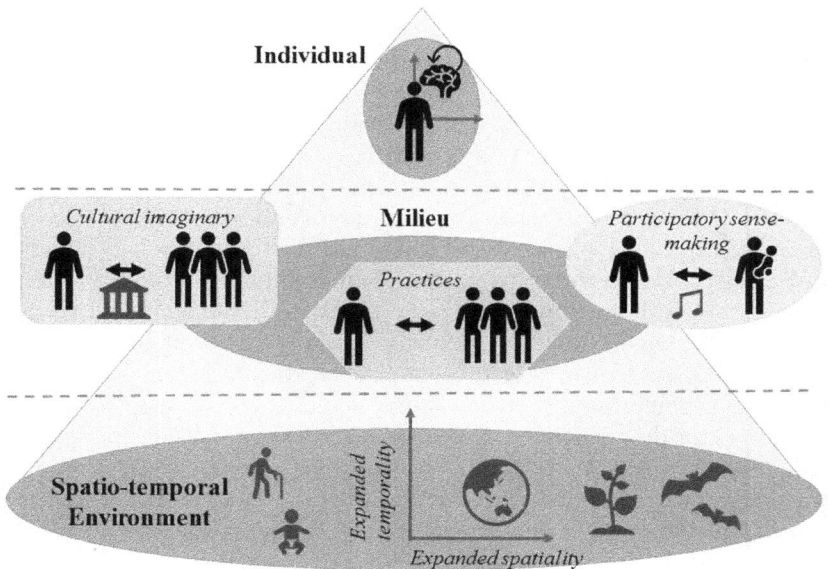

Figure 2.1 Diagram of the model environment-milieu-individual

involving humans is ethically loaded. In Figure 2.1, the highest level of the circular pyramid is the individual human: the phenomenological agent capable of ethical decision-making. From the phenomenological perspective, it is the place of thinking, feeling, perceiving, introspection, self-criticism and self-reflection. From the perspective of the observer, other elements intervene at this scale of analysis, such as cognitive biases, neuroscientific knowledge and perception-action couplings, hence the symbol of the brain and the circular arrow. When we think, alone, about how we ought to act, who we want and should be and why, it happens at this level of "morality".

Yet, we are never really "alone" when we think and reason about morality – or about anything else. We are thinking with words, concepts, images and ideas that pertain to our sociocultural milieu. We are borrowing the tools for our personal self-reflection from the common toolbox of the cultural backgrounds we have access to. We are making sense of the world together with other human beings (hence participatory sense-making, see 2.3.1). Everything we can imagine is also composed of elements that we found in the existing material world and immaterial cultural imaginary. A unicorn is simply a composition of a horn and a horse, two already existing elements. Imagined combinations of ideas are limited because they are borrowed to a context that pre-constrains them, and the means of the agent doing the "bricolage" are not defined by a project or a goal, but by their availability (Lévi-Strauss 2008). Finally, the ways in which we act are, even when we are alone and nobody is looking at us, informed by our cultural norms and practices.

The intermediate level of the milieu encompasses ethical *relations* between humans, and between human individual and otherness, including nonhuman living beings and the spatiotemporal environment. It is the level of human understood as relational and "in group". Note that the human individual can never be artificially extracted from this medial level ("medial" being the adjective of the word "milieu"). Neither can she extract herself from the environmental level.[2] There is no human existence *out* of these two other levels. If we flip the situation upside down, we can also notice that we have *no direct access* to the level of the individual alone, nor to the level of the environment without human beings, because all the tools we use to think about them and the experiences we have of them are mediated by the medial level.

Before going into more details in the exploration of the milieu, I need to clarify what conception of the self such a perspective entails. To better approach to this complex question of what the human self is, I want first to distinguish two distinct perspectives: the external observer perspective and the internal phenomenological perspective of the agent.

2.1.2 Observer versus phenomenological agent

An object can appear very differently depending on the standpoint from which we are looking at it. This is no different with concepts, especially for ideas that are heavily normatively loaded such as ethics, the good and happiness. Sciences are supposed to take as much as possible the standpoint of the neutral objective observer. They look at facts and relations that are visible from the outside and try to grasp them in the most objective way, that is, as we discussed previously about rationalism, the most consensual way. In contrast, phenomenology seeks to analyse, express and understand the experiences or these ideas, facts and relations from the inside.

There is no clear-cut border between the perspective of the phenomenological agent and that of the observer in part because any observer is also a phenomenological agent, and because any phenomenological agent borrows knowledge from the observer standpoint and can be the observer of her own phenomenological experiences. Both standpoints are permeable to each other. Still, when discussing ethical matters, making it clear from which standpoint one argues can help to avoid misunderstandings. The observer perspective tends to be situated outside of the specific circumstances of the event or idea to be analysed. This perspective is partially detached from the spatiotemporal standpoint. At least, it is detached from the everyday life preoccupations, including aging and dying. In contrast, most ethical decisions and actions are entangled in everyday life worries and embodied in a specific aging and evolving body.

An example of a philosophical debate steeped in this ambiguity between perspectives is the philosophy of the "good life". One popular family of theories on what a "good life" consists of focuses on the set of actions and events composing the life of an individual and judges its consistency, morality, integrity and success. This observer perspective on one's life faces the problem that to carry a judgement on one's life, we should wait for this individual's death. That is why a highly morally negative act committed at the end of one's life could affect the general assessment. From this viewpoint, assessing the goodness of one's life can be done only *after death*, by an external, supposedly neutral observer. This view conflicts with the intuition that a good life depends on what the individual wants and decides to do, which might change depending on external circumstances of one's life. Indeed, instead of focusing on an "objective" assessment post-mortem when the individual cannot express opinions anymore, would not it be more constructive to focus on the day-to-day struggles of this individual with specific ethical dilemmas in her life while she can still act on them?

This debate partly echoes an important distinction in ethics between Kantian deontology and consequentialism. For a proponent of an exclusively consequentialist theory of ethics, what a good life is can be judged only on the consequences of one's actions, regardless of one's intentions. An observer view on one's life is thus sufficient to assess the goodness of one's life. On the contrary, a deontological approach necessitates to know the intentions and moral beliefs of the individual in order to carry a judgement on one's actions. A deontological perspective thus requires to include an inquiry into the phenomenology of the agent. As any observer is also a phenomenological agent, she observes the other and the world through her own phenomenological glasses, even when she makes her best effort to be objective and external to her own prejudices and values. Moreover, these distinctions regarding what a good life consists of from the perspective of the phenomenological agent or of the observer are related to the conceptions of the human self they presume.

2.1.3 The construction of the human self

It is hard to explore *what* the human self is without bouncing into normative questions such as what a good life is. We can start with the apparently easier question of *where* the human self is. The self is always anchored in one body situated in a specific spatiotemporal point in the world. The embodied self seems to be in between perception and action. From the perspective of the observer, we first see others' bodies that are already covered by indications of their identities, such as clothing, tools and body language. From the perspective of the phenomenological agent, one perceives the world as composed of different possibilities for action. James J. Gibson describes these possibilities as affordances: "The affordances of the environment are what it offers the animal, what it provides or furnishes, either for good or ill. [...] It implies the complementarity of the animal and the environment" (Gibson 1979). Moreover, we do not consciously think about most of our daily actions (breathing, walking, etc.). We perform them automatically and very efficiently.

How do we think? This is the question where the explanatory gap between the perspectives of the observer and the phenomenological agent is the most apparent. In recent years, progresses in neuroscience and cognitive sciences are giving us more information from an observer standpoint, shedding light on cognitive biases and on the limitations of our thinking (Northoff 2014; Wang et al. 2018). The problem of the consciousness, impenetrable from outside of the phenomenological agent position remains. Only oneself knows how she thinks. But we can try to communicate with each other about what is happening in our heads by exchanging descriptions and narratives of our thinking. These narratives are using words or ways of expression loaded with cultural connotations, social expectations and normative values. The listener to this narrative is also understanding it through her own glasses. Moreover, one will tell a slightly different story depending on who is listening to her by assessing the other's values and expectations and adapting her own narrative correspondingly.

This leads us to the question of *which* self? We all have multiple social roles, as a mother, a sister, a citizen, a teacher, a woman, a native of a region, etc. We are pushing some of these social roles forward in a specific interaction with someone else. We sometimes want to be *seen only as* one specific social role. From the observer perspective, we can watch these differences in attitudes, body languages and narratives about oneself (Heberlein 2012). We are telling a different story of who we are to different people. From the phenomenological agent perspective, self-reflection is a form of debate between different voices inside our heads. Our own self is multi-voiced. Bring these elements together, and what we refer to as our self is composed by multiple voices rooted in slightly different experiences, emotions, worldviews and values that might oppose each other. What is unifying these voices is that they are all appearing from an embodied act of consciousness. The narratives that we build to describe ourselves and to justify our decisions and actions are often not coherent. They can be composed of episodes without a logical narrative linking them (Strawson 2015). Succession of events in lives is usually accidental or coincidental, and surely not always filled with clear goal-reaching continuity (Lamarque 2007).

The construction that each of us builds to make sense of the succession of events of consciousness, memories and future aspirations matters because it is used as a normative justification for actions. It involves understandings of what is good for oneself and how one *should* behave. Yet, this construction is not done in a vacuum "alone", but in constant dialogue with others and the world. Some philosophers describe this construction as "a discursive action and performance relationally embedded" (Gergen 2011). From this perspective, the self is dynamic and "intrinsically uncertain because it is never complete at any moment in time and is always in need of a different part of itself, in order to arrive at some clarity in its relation to itself and the world" (Hermans 2011, 674).

In my three-level model, the individual self is always situated in a socio-cultural milieu. The introspection the self conducts at the level of morality is always informed by collective voices, ideas, worldviews and theories coming from the level of ethics. The way of thinking and reasoning is largely determined by the theory of rationality that is consensual in our milieu. Our personal reflections are reinforced by values and theories that are socially shared. As Neisser writes, "our assumptions about ourselves form a "web of beliefs", drawing their meanings from one another and providing each other with mutual support" (Neisser 1991, 198). Most of these assumptions, beliefs and meanings about oneself are coming from the milieus we grew up in and we interacted with. Thus, one's web of beliefs about oneself is not isolated within the individual thinking, but is reinforced and continuously readapted through interactions with others and the world.

2.2 Background of the concept of milieu

What is the milieu? The word "milieu" appeared in English in the middle of the 19th century from French and refers to "a person's social environment" (*Oxford English Dictionary* 2000), or "the people, physical, and social conditions and events that provide the environment in which someone acts or lives" ('Cambridge Dictionary | English Dictionary, Translations & Thesaurus' n.d.). Literally, in French, it can mean the middle or centre of a place ("lieu"), but also everything that surrounds this central point. It can be the centre of a target or a circle. At the same time, it can refer to everything surrounding a being, in the middle of which this being exists. Etymologically, the word "milieu" comes from the Latin "medius locus", "middle place" (Ménage 1650).

Because this book aims at discussing human decision-making and actions, we focus on human beings at the centre of the milieu. From the observer perspective, the individual exists in the middle of her milieu. She is influenced by her sociocultural and environmental surroundings directly and constantly. She also dynamically reacts to events occurring around her in a certain way partially determined by her specific sociocultural backgrounds. She is one nod in the wide and dynamic web of interrelations, situated at a specific place and time. From the perspective of the phenomenological agent, she generally does not see her surroundings as a milieu, a specific environment shaped by sociocultural dynamics. She sees her surroundings as her reality. She sees only the milieu and can hardly access the world without passing through it. Entangled in these multiple relations, she is also an agent, acting on them. Her everyday life actions support and change the web of relations composing her world.

Yet, most of the time, the word "milieu" is used from a collective approach, in relation with "cultures". On the one hand, humans shape the world and leave their traces on it. On the other hand, humans are shaped by the milieu. The milieu is the mould of culture, and culture concomitantly makes the mould. In Berque's words (Berque 1996, 2000), the milieu is thus at the same time the imprint and the matrix of humans beings. The word matrix originates from the Latin "mater", mother. It is that from which something originates and develops. We humans are born and grow up in particular medial matrices. In turn, imprint refers to the traces that we leave in our surroundings. We will now explore the literature around the philosophical concept of milieu. Then, we will delve into both sides of this coin: the milieu as a historical matrix and the milieu as a collective imprint.

2.2.1 Watsuji's fūdo

The term of milieu gained popularity in the middle of the 19th century in the movement of the possibilism, itself developed in opposition to natural determinism. For Paul Vidal de la Blache (1845–1918), the main proponent of possibilism, the one same milieu can be used in diverse ways, depending on production technologies that support diverse ways of life (Vidal de la Blache 1903). During the same period, Jakob von Uexküll, a pioneer in the field of ecology, introduced the distinction between *Umwelt* and *Umgebung*. *Umwelt* refers to the environment of an animal from its own perspective, a self-centred world. In the case of human individuals, this is the perspective of the phenomenological agent (Von Uexküll 1909). On the contrary, the *Umgebung* is an *Umwelt* seen by another observer. In other words, for Uexküll, each species has its own universe determined by what they can and need to do, covered by corresponding meanings. In recent terminology, each species is perceiving and using its environment according to the affordances (Gibson 1979) depending on their own embodied possibilities.

Even today, in place of the ambiguous concepts of nature or world, sciences and most of mainstream international political discourses prefer to use the supposedly more precise word "environment", in Uexküll's sense of *Umgebung*, as the natural world surrounding us seen from the perspective of an observer. The perceiver is erased, and the "natural environment" is taken to be independent and isolated. The idea of "ecosystem services" attempts to illustrate the various ways in which we, humans, benefit from the natural environment. But it generally fails to recognize the fact that we are precisely the ones designing and defining this independent object standing apparently on its own in front of us. We are then left with an illusion of objectivity precisely where the diversity of worldviews reaches its peak; in our relationship with the world.

Almost a century ago, the Japanese philosopher Watsuji Tetsurō (1889–1960) pointed to this elusive relationship between humans and nature with his idea of *fūdo*. Watsuji was professor of ethics at Kyoto University (1925), and then held the chair of ethics at Tokyo University from 1934 to 1949. The concept of milieu developed in my book is directly inspired by Watsuji's idea of *fūdo* – which was translated into French as milieu (Watsuji 2011). The word *fūdo* (風土) is composed of the Chinese character for the wind and the character for the soil. It attempts to grasp the relation between humans and their environment while recognizing the essential subjectivity of the relation itself. Watsuji writes in the first lines of his book titled *Fūdo*:

> What we usually think as the natural environment is a thing that has been taken out of its concrete ground, the human milieu-ity, to be objectified. When we think of the relation between this thing and human life, the relation itself is already objectified. This position thus leads us to examine the relation between two objects; it does not concern human existence in its subjectivity. On the contrary, this subjectivity is what matters in our opinion. Even if medial phenomenon is here constantly questioned, it is as expressions of human existence in its subjectivity, not as the natural environment.
>
> (Watsuji 2004, 1; 2011, own translation)

"Human existence in its subjectivity" also is what matters to us, because it is the keystone of environmental ethics. Interestingly, the first lines of *Environmental Values* (O'Neill, Holland, and Light 2008) strikingly echoes Watsuji's considerations, without mentioning him:

> There is no such *thing* as the environment. *The* environment – singular – does not exist. In its basic sense to talk of the environment is to talk of the environs or surroundings *of* some person, being or community. To talk

of the environment is always elliptical: it is always possible to ask 'whose environment?'

For Watsuji and in my view, the natural environment as seen and inhabited by subjective phenomenological agents is the milieu. It is in this subjectivity that rests normative assessments about good life and ethical actions. From the standpoint of a subjective human, the "environment" appears as a web of significations and symbols. The milieu is subjectively perceived and co-created by humans. For Watsuji, we are constantly in dynamic cycles of codetermination with the milieu. Hence, like the movements of a restless pendulum, we negate our self to identify with our milieu, and then we assume our independence and difference from the milieu to come back to our self. For an individual, this swinging only stops with death.

This dynamic and codetermining relation between the self and its milieu entails a conception of the self that is relational, ever-changing and adaptive. Watsuji's own conception of the self has been shown to be relational (McCarthy 2010; Johnson 2019). Not only is our identity built in relation with others, but so are our ways of life, ideas and practices (Sevilla 2018, 23–34). Even at the bare physical level, we keep exchanging with our surroundings, through breathing, eating, etc. (Yuasa 1987). For example, recent research about the human microbiome shows how we keep exchanging bacteria with our surroundings, and how these influence our health and our mood (O'Doherty, Virani, and Wilcox 2016). As microbial communities seem to be interconnected around the globe, we could say that we are interconnected at the planetary level through cyclic flows of microorganisms (Mestre and Höfer 2020).

In brief, the individual cannot be abstracted from the milieu and the environment. We do not construct ourselves in a vacuum (Gergen 2011), but intertwined in multiple relations with others and with the milieu. This conception of the self fits with the former constructivist description of a dialogical and dynamic self. The self understood as a web of mutually supporting beliefs is constantly challenged and updated with new knowledge concerning its surroundings and itself. It also constantly confronts itself with the webs of significations and symbols embedded in the milieu. Both consciously and automatically, we continuously make comparisons between the webs of beliefs of our self and of our milieu. In case of important mismatch, the difference may spring to our attention and force a reassessment of our own beliefs. Widespread mismatches even may hinder our capacity to thrive in this specific milieu.

Strikingly, our relationship with the world, namely the milieu encompassing the natural environment, mirrors the construction of our identity. Obliterating the former under a veil of apparent consensual objectivity hides the various nuances and differences that make us who we are as individual agents and members of sociocultural groups. Cultural identities are visible in the milieu. The milieu itself is always lived and codetermined, criss-crossed by our practices and usages of the space. Practices and usages that are proper in a specific space are largely determined by cultural norms and expectations. As such, the milieu is also a place of intersubjectivity. It is a place where people meet, and on which people project representations, significations and symbols through their common imaginary.

As a place of interactions, the milieu is charged with norms regulating how we should exchange, speak and behave with other humans and with other elements of the milieu. As it appears clearly in the three-level model, the milieu is also the place of ethics. Ethics emerge from the relations between humans among themselves and with their milieus. But my account of Watsuji's milieu would be incomplete without mentioning his concept of betweenness (*aidagara* 間柄). For Watsuji, humans are relational *by definition*, for the second Chinese character of the Japanese word "human" means betweenness and relation. Being human is "being inside the world" (*yo no naka* 世の中), which means not only taking a place within the social web of relationships, but also being a part of the environment.

This "space-in-between" the milieu and individuals is what makes actions and self-reflection possible. We can think and act precisely because we can distance ourselves from our milieu. We distance ourselves from the milieu briefly, temporarily and partially, in the midst of the constant codetermining cycles, and anchored in our embodiment, but long enough to be (partially) autonomous agents. That is, long enough for ethics to emerge from this betweenness, as we make our own choices and we take actions as reflective agents. On top of this relationality, agency is another essential aspect of what it is to be human.

Betweenness is also at the heart of the contradiction contained in the concept of milieu. Watsuji notes that we cannot presuppose neither the individuals before the relation, nor the relation before the individuals. Indeed, neither has "precedence". This applies to all expressions of betweenness such as words, facial expressions, ways of living, customs and ethics: They do not exist prior to individuals, nor do individuals exist prior to these "moments", which constitute betweenness. Watsuji recognizes this "contradictory relationship" in the movement of emergence of the individual and betweenness (Watsuji 2007, 1:58).

This contradictory aspect echoes the bivalence of the relation between humans and their milieu, in that the milieu is both the result of humans' ways of life, and what makes human life possible. This bivalence is captured by Augustin Berque's application of the bivalence imprint-matrix to the idea of milieu. Berque first translated Watsuji's book *Fūdo* (1935) into French and Watsuji's concept of *fūdo* as milieu (Watsuji 2011). He developed his own idea of milieu around the concept of *fūdo*. He traces the idea of the milieu back to Plato (Berque 2000). In Ancient Greek, the *chora* (Χώρα) refers to the countryside surrounding and subsiding the city. In Modern Greek, it refers to both the territory and the village. In *Timaeus*, Plato compares the *chora* to both a mother and an imprint (Plato, *Timaeus*, 50, d2, c1). He recognizes the contradiction in the concept, namely that the same thing designs simultaneously the centre and the surroundings, but insists on its existence. Berque notes that like the milieu, the chora is at the same time a matrix and an imprint. Berque relates these apparently self-contradictory concepts of *chora* and milieu to the idea of the landscape. Indeed, as human geographers have been discussing for decades, the landscape seems to be at the same time existing "inside" the social perception and imaginary of a group, and "outside" of the human world, as it is composed of mountains, seas, trees and rocks (Besse 2018).

Watsuji includes the landscape within his idea of milieu, as he writes in the first line of the first chapter of *Fūdo* that *fūdo* is a "general term including for a particular region, climate, weather, geology, soil quality, geographical features, landscape *(keikan)* etc." (*Fūdo*, 1935). But Watsuji did not use the matrix-imprint bivalence in his description of the milieu. Yet, interestingly, he mentions briefly the word "imprint" when he discusses landscape to warn us against the risk of opposing human activities and a passive natural environment. He writes:

> Thus, the landscape of a region cannot be an imprint carved by the group living in this region by itself on a nature standing opposed to the group. Instead, the group expresses the content of the group's ethical organization into the shape of the land. Then, landscape is an internal scene in the human existence; it is not an environment which surrounds from the outside human beings.
>
> (Watsuji 2007, 315–16, emphasis in the original, own translation)

In other words, we should not misunderstand the relation between the group and the milieu as if the landscape were the result of groups "trying to make the shadow of themselves cast on a particular nature last forever". Watsuji insists that far from being a given, "the environment or natural region" is changed by human activities throughout history (Watsuji 2007, 316–17). In this "environment or nature", there are "cultivated countryside, planted mountain areas, towns lined with water streams, and villages scattered across the plains", which are nothing but the "shape" of the human groups. It questions the very possibility of isolating something supposedly "purely natural" deprived from any traces of human activities (Droz 2021).

Still, Watsuji continues noting that it is possible to approach the alterity and outside character of the concept of environment by an abstraction process. For example, natural sciences attempt to reach the neutral reality of environmental phenomena by minimizing as much as possible interferences from human experience. Indeed, while we always grasp the world through the intersubjective filter of the milieu coloured by the cultural context, the quest of natural sciences can be understood as to avoid cultural relativism by constructing another mode of intersubjectivity. Watsuji warns that we should not forget two things. On the one hand, this environment is the result of the "expressions of human existence", that is the side of the imprint. And on the other hand, we should not forget that the "subjectivity of human existence can only be grasped from within" the landscape and the environment, which echoes the matrix side of the milieu. Pairing the idea of imprint with the idea of matrix, like Berque suggested, avoids the risk of objectifying the environment and abstracting a subject-object relation. In other words, milieus are both the matrix and the imprint of human beings living on Earth.

2.2.2 Collective imprint

"Human imprints on Earth are collective". This sentence can be understood in two ways. First, human imprints are collective in the sense that they are the product of societies and cultures. Second, human imprints are collective in that individual actions themselves are collective. The first sense is the most common in most of Watsuji and Berque's works. This understanding faces the difficulty to demarcate what a society and a culture is. Both authors leave this question vague, drawing on the common sense of the reader. Yet, this vague common sense is problematic, insofar as it rests on the assumption shared by the dominant majority of a community (in this case, the readership). The delimitations of the group counting as society or a culture, too, are highly normative and political. Indeed, delimiting what a culture encompasses inevitably requires to mark out some elements and individuals. Deciding who is included in and who is excluded from a group is always a political and ethical decision that has important consequences in the real world. Despite these difficulties, it is impossible to ignore this question when discussing real-world problems.

The interpretation of milieus as the imprints of societies and cultures is widespread in the literature. To seek slightly less ambiguity, I define here collective imprint as the whole of the consequences of the actions (and inaction) of the individuals that are part of a specific group. For each real-life usage, it is necessary to state what a specific group we are referring to, and assume the highly probable political implications of such a statement. To use the type-token distinction, it is the type of humans to leave imprints on the world, but each group's imprint is a specific token of this relationship.

Then, how are individual actions themselves collective? The actions of an individual agent are collective because they rest on collective significations and values, and more concretely, on tools and usages pertaining to a specific cultural group. This can be observed at the basic level of the experiences that an individual has and at the level of the concrete technical and economical relationships on which individuals are building their sense-making. I borrow two examples to illustrate this point: Watsuji's example of the coldness and Berque's example of the pencil.

As I mentioned, the Japanese word *fūdo* is composed by the Chinese characters for wind and for soil. To feel and perceive the dynamicity of winds, one needs to be anchored and situated at a place on soil. The soil is where one is standing, one's point of observation and perception of the caresses of the surrounding air. Watsuji's example of the cold makes the importance of the perspective of the phenomenological agent clear. Coldness does not exist to an observer. A neutral observer can measure the temperature but cannot know if it *feels* cold or not for the living being experiencing it. Still, coldness is not purely internal and individual, it is a shared experience insofar as it is defined with others through words, salutations, social activities, etc. My Indonesian friend sitting next to me does not experience the same coldness as I do, because what she grew up with being defined as cold is radically different from my upbringing. To her, coldness is exceptional and linked to an imaginary of pictures of foreign landscapes seen from inside her home, through media. To me, coldness is the sign of wearing winter clothes, starting a fire and eat warm meals. Practices and norms that we learnt from the milieu we grew up in are directly influencing our experiences. Our cultures have been built around the relation between communities and milieus. In Watsuji's terms, coldness exists within the relation, the betweenness (Watsuji 2004, 13). Watsuji follows the possibilist paradigm and argues that facing similar environmental constraints, different individuals and cultures will elaborate diverse solutions.

The milieu is also a set of eco-techno-symbolic relationships. Besides norms regulating human interactions, the milieu is designed according to specific usages of elements of the natural environment. Concrete practices associated with specific symbols and meanings are linked with corresponding technologies. Tools are key elements of the milieu that clearly reflect the specific relation between a group of humans and their milieu. Berque uses the example of the existence of the pencil as intrinsically relational (Berque 2000, 92). The pencil is part of a first set of symbolic systems: writing, words and languages, which implies human relationships, sense-making and the cultural imaginary. Then, it is also part of a second technical system: the forest and trees cut to produce the wood, the machines used to process it, the paper mills producing the paper, etc. These activities all are imprints that constitute the milieu.

The milieu is shaped by our usages. Landscapes are striking expressions of these relations. One look at Indonesian terrasses rice fields in contrast with Swiss mountainous wooden pastures is giving us information about the differences of the concrete relations that these two cultures have developed with their specific milieus. Usages are informed not only by collective representations of the milieu, but also by the accessibility of techniques and by traditions. The diversity of ways in which humans built and designed their surroundings are well known and widely discussed not only in geography, but also in engineering, urbanism and other applied sciences. For instance, the impact of irrigation techniques on the organization of the space and societies is not to be demonstrated anymore.

Most of current environmental problems are also collective imprints. In some cases of locally delimited soil, air and water pollution, the borders of the group that contributed to this imprint can be marked. Yet, pointing out the contributory group is much harder and even almost impossible for most environmental problems, as the most responsible people often are not the ones living where most of the pollution is concentrated. Climate change and sea-level rise are also human collective imprints on Earth, but their responsibility cannot be assigned to one, or even to a set of sociocultural groups. These global medial imprints are results of human activities. Moreover, the human groups involved in these imprints are not only the currently existent ones, but also the past generations. Cultures and societies today rest upon their respective and common histories.

2.2.3 Historical matrix

The milieu is historical. The milieu carries the signifying traces of the past, co-creates significance with the living humans and transmits it to the future generations. The organization of a particular medial matrix, both in terms of social contracts and in terms of usage of the land, is the fruit of the ways of life of past generations that inhabited the place, and of their interactions with travellers and other groups. This process unfolds over time, and that is what Watsuji refers to as historicity. He notes that "we live with the past human beings within us", and that this "rich past content without limits that defines the present" is the source of the present "unity of the ethical social organization" (Watsuji, 2007, 156, own translation). For Watsuji, the social organization is multilayered, spanning from the scales of the family, the village, up to the scale of the nation, and ethics emerges from the relations that bind together the members of these communities (Carter 1996).

The signifying traces take different forms, starting with the tangible heritage. Tangible heritage consists of buildings and artefacts recognized as having important cultural significance. The Great Wall of China is an example, as well as Roman roads, sites of the Industrial Revolution, etc. We can also enlarge the definition of tangible heritage to include any human-made material element carrying important symbols and significations. Of course, the decision of what is important is determined by normative values and opened to debates and negotiations. Regardless of the official listings such as the World Heritage List of the UNESCO Heritage Center, the point is that the milieu is literally covered with historical heritages entangled in their environmental settings. The heritage can also be intangible, referring to music, tales and dances, and described in UNESCO's Immaterial Heritage Convention (Labadi 2012, 127–55). These can be recorded in material things such as books and videotapes. But they all need to be performed, because their survival depends on the performers and the audiences. This also includes craftsmanship, and techniques of using some natural elements as tools and materials to make something else. The knowledge of weaving robes and cultivating certain crops is intangible, but still a central element of the design of milieus.

Traditions are also signifying traces of the past. They are the inherited and established patterns of thoughts and practices. The very fact that some traditions are not verifiable and that many have been recently constructed from scratches reveals the need that societies have for a common imaginary supporting practices and guiding behaviours. Celebrations and festivals give rhythm and meaning to life events such as birthdays and weddings, and ground feelings of belonging and togetherness within a community.

These traditions are important features of the common imaginary. The common imaginary is the set of values, meanings, stories and concepts that are shared by a particular group. It is often largely influenced by the linguistic community and overlaps with religious backgrounds. It includes floating and unspoken social and ethical norms some of which are crystallized into laws. Being written, recorded and debated, laws are easier to analyse and give an interesting access to common imaginaries of older times and faraway places. Norms and laws are normative expressions of worldviews through practices. Because they are often the product of long historical processes of negotiations, these concrete structural expressions of conceptual worldviews retain a strong inertia and generally do not change easily. In the case of environmental issues to be tackled with rapid societal changes, the inertia of laws can be an obstacle. Yet, it also gives us precious information on past and present ethical worldviews, on the shoulders of which we are continuing to build normative systems.

Finally, languages, symbols and meanings are essential parts of the historical medial matrix. One crucial element of this matrix is the linguistic system. We learn our mother tongue(s) not only from our caregivers, but also from people surrounding us, and from symbols, sounds and colours that cover our surroundings. Since infancy, we absorb all this information and make it our own, as an essential part of our identity. Belonging to and coming from a particular sociocultural community, that is supported by a medial matrix, is definitely a key element of our identity. But this aspect of individual identity goes beyond individual life. As members of a particular sociocultural community and children born out of a particular historical medial matrix, we also carry the signifying traces of the past and co-create significance in the present.

Watsuji highlighted the importance of the historicity of the milieu. Historicity unfolds through every human action, considered as adapted reproductions of ancestral gestures characteristic of the belonging to a sociocultural community. More importantly, for Watsuji, diverse forms of human communities developed as systems of dynamic movements of innovations, that is, as the expressions of partial and temporary freedom from their natural environment. Every human community is grounded on the spatiotemporal structure of human existence. Finally, this structure is defined by the cyclic pendula movements of negation of the self to identify with the milieu – a dissolution of the self into the other, the community or the milieu, because the collapse of difference leads to the negation of the self – and of negation of the negation – of the social or medial – by coming back to the agency of the self (Droz 2018, 149–50).

> No social structure is possible if not grounded on the spatial structure of subjective human beings; and temporality, if it is not grounded on social existence, cannot become historicity. [...] The individual dies, the relation between individuals changes, but while dying and changing, individuals live and their betweenness continues. [...] What is "being towards death" from the perspective of the individual is "being towards life" from the perspective of the society.
>
> (Watsuji 2004, 19–20)

Historicity is to temporality what the milieu is to spatiality: its humanely subjective relata. Watsuji draws a clear line between human individual existence characterized by death and the social existence of the community. The social community survives the individual. This makes sense if we understand the social community as the whole composed by the cultural imaginary, meanings, norms and practices characterizing a specific way of life. Yet, the individual does not die without influencing these by sustaining them and changing them.

Nevertheless, we face here again the problem of drawing the borders of the community. We can hardly identify a social community from within. Usually, we can identify it by contrasting it with others and by highlighting the differences. In his book on Watsuji's work titled "Ethics", Anton Sevilla refers to this aspect as the "self-realization of identity through difference" (Sevilla 2017, 73). He notes that for Watsuji, a community (or in Watsuji's problematic words, a "totality") can "only realize its own identity in history by coming into relation (or conflict)" with other communities. In other words, a community can only fully realize the particularities of its medial matrix in relation with others (Bath 1969). From this perspective, external diversity seems to be an essential feature of – maybe even a necessary condition for – internal identity. Possibly, the very fact that we can relate to the world and to ourselves in a rich diversity of ways is precisely what makes us human, as communities and as individuals.

2.3 Updating the idea of milieu

To use the idea of milieu in contemporary global environmental ethics, some more developments are necessary. Updating the idea of milieu to apply it to the purpose of my book led to the elaboration of the conceptual framework of the milieu. This framework provides the basis for the discussion of sustainability and responsibility in the next chapters. I also highlight some characteristics of my conception of the milieu that will regularly appear and lead to important conclusions in the next chapters. Finally, I position my conception of the milieu in comparison to other rival ideas in the field of environmental ethics.

2.3.1 The conceptual framework of the milieu

Watsuji was a traveller, and a strong point of his concept of milieu is that it recognizes the diversity of cultures anchored in different milieus, and how these mutually shape each other (Ota 2018). He also insists on the dynamicity of the cycles of codetermination between the self and the milieu, rejecting natural determinism. Nevertheless, like most of the past philosophers, Watsuji seems to assume that cultures are directly linked to bounded communities grounded in a particular land. For our contemporary use in this chapter, this is an important limitation of his work, along with the absence of mention of responsibility towards the milieu and of internal mechanisms of oppression and domination within communities. In today's globalized context, any ethic aiming at assigning responsibility for environmental harm needs to address these issues. Moreover, in order to assign individual moral responsibility, we must distinguish clearly between the individual phenomenological agent and the community, and how they relate to each other within their relation with the milieu.

Watsuji's philosophical perspective crystallized in the idea of milieu is characterized by its rejection of an abstract and objectified natural environment. This rejection is accompanied by a warning that sciences' tendency to over-objectification misses an essential aspect of human existence, namely, the phenomenological perspective of human existence and its concrete and situated nature. Finally, Watsuji's philosophy places ethics at the centre of the relation. Some scholars argue that the place of ethics in Watsuji's philosophy lies in the mutual relations between human beings (through *aidagara*) (Liederbach 2001). Setting aside the philological debates around the precise interpretation of Watsuji's *aidagara*, my conception of the milieu places ethics not only within the relations between human beings, but also within the relations between human beings and their milieu. As the milieu is the environment surrounding a phenomenological agent that lives in it and perceives it in a particular way, the relation between human beings and their milieu includes the impacts on the environment by human activities that are judged harmful by human beings. The objective of my book is not to provide an interpretation of Watsuji's philosophy as faithfully and close to the text as possible, but to take inspiration from his ideas and adapt them for contemporary applications. In order to apply the idea of milieu to contemporary environmental ethics, my conceptual framework of the milieu takes over three aspects from Watsuji's philosophy, namely, the criticism of an abstracted environment, the insistence on the phenomenological perspective and the place of ethics within the relation.

Drawing on and moving beyond Watsuji Tetsurō's idea of milieu and Augustin Berque's concept of matrix-imprint, I develop a conceptual framework that clarifies the cyclic relationships bounding together the phenomenological individual agent, the medial imprint, the community and the medial matrix. The process unfolds as follows (Figure 2.2). The community shapes, more or less deliberately, the surrounding environment through habits and repeated practices over time, thus creating a medial matrix. In turn, the milieu as a matrix informs, guides and constrains the practices of its inhabitants. Individual phenomenological agents live constrained, guided and inspired by the web of norms and meanings of their milieu. Within this medial matrix, the individual agent retains a relative flexibility to make reflexive ethical choices. By acting, the individual agent then leaves medial imprints on the milieu. Finally, the whole of the individual imprints of the inhabitants of the milieu shapes and changes the milieu. The milieu is maintained and changed through the intertwined imprints of multiple individual agents interacting with each other and with the milieu. Figure 2.2 shows the cyclicity of these medial processes.

The conceptual framework of the milieu allows to distinguish the perspective of the phenomenological agent who inherits of a particular sociocultural milieu which roots her identity and informs her practices, from the intergenerational collective historical processes that contribute to building the medial matrix. The milieu is continuously changing through historical processes that unfold across generations and compose the norms, usages and meanings in which human beings live in the present. The historical collective processes shape the medial matrix through medial imprints. Imprints are understood here as *all* the traces left on the milieu by the ways of life of a group or an individual. This wide-encompassing concept includes traces left non-deliberately by habits and omissions.

The medial processes around the phenomenological agent and the community are circular. The medial matrix surrounding an individual is shaped through a historical process involving the community. What counts as a community in our framework of the milieu is deliberately kept vague, as

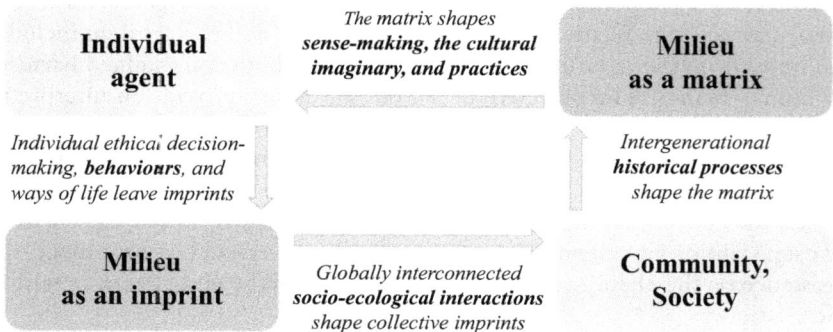

Figure 2.2 The conceptual framework of the milieu

it takes very different shapes and forms depending on the case to which the framework is applied. Still, insofar as the historical processes creating the medial matrix are involving multiple people over countless generations, they are "collective". An inclusive view could interpret the community as a multispecies community that includes the living beings – human and other-than-human – who share and inhabit a particular milieu. A more conservative view could restrict the community to the people who inhabit a specific milieu. The literature on the milieu tends to assume the community to be exclusively human and to keep what counts as this "human community" ambiguous. It also refrains from differentiating individuals from the groups. Yet, to place individual moral responsibility and to apply the framework to concrete questions, such a distinction is necessary. The framework of the milieu distinguishes between the collective aspects of the imprint that result from negotiations of practices between people over long periods of times, and the perspective of the phenomenological agent.

The milieu is at the interface between the subjectivity of the individual phenomenological agent and the communality of the society or group that shares and influences it. Insofar as the milieu is lived and perceived, it is always so by an individual. Yet, the ways this individual lives and perceives her surroundings are not exclusively individual. Instead, they are largely shared and communally designed by the community through historical collective processes. In other words, the subjective milieus of several individuals in a single community are likely to be highly similar and largely overlapping. We can then speak of the milieu of the community in the singular form. Further, different communities are influenced by and design different milieus in the plural form. When travelling, an individual encounters and is influenced by different milieus. Milieus are diverse interpersonally, as each individual lives and experiences her surroundings in an irreducible personal way, even if this way is only slightly different from her peers' ways. But more importantly, milieus are socioculturally and ecologically diverse insofar as each community organizes their surroundings in particular ways through interactions with the natural environment. Finally, from the perspective of the phenomenological agent who travels and uses different milieus, milieus can appear to be intra-personally diverse, because one individual can carry and be shaped by the influences from several milieus. Thus, the plural form of milieus is used to stress this multilayered diversity, primarily the sociocultural and ecological diversity of milieus.

2.3.2 Characteristics of the milieu

In the framework of the milieu, there are four main characteristics (Table 2.1) of the concept of milieu that distinguish it from the rival concept of environment: Milieus are (1) built in mutual cyclic relationship, (2) experienced by phenomenological agents, (3) connected globally in a dynamic network and (4) continuously changing. Each of these four main characteristics echoes the four aspects of the conceptual framework of the milieu sketched in Figure 2.2.

First, the concept of milieu highlights the *mutual cyclic relationship* binding together individual human beings, communities and their surroundings, including the natural environment, ecosystems and other species. This first characteristic appears clearly in our description of the processes of the medial matrix shaping individual phenomenological agents, especially when we approach these processes from the observer perspective. It rejects the common connotation of the environment as an over-objectified passive receptacle. On the contrary to the scientific paradigm that usually objectifies the environment as a passive receptacle independent from humanity, the concept of milieu recognizes the *inseparability* of humans and their environment, as humans and their milieu co-defined each other. Then, the concept of milieu can prevent utilitarian analyses blinded by cost-efficiency calculations by integrating from the very start the subjectivity of ethical values and usages of the place. Besides, it does not exclude the fact that some milieus might advocate a strong dualism between humans and nature; this dualist perspective is understood as a locally anchored worldview, not as a universal objective truth.

Second, the concept of the milieu is characterized by the emphasis on the deep *phenomenological experience* of the individual as the ground for external ethical action on the milieu. In contrast, the environment is usually taken to be neutrally and objectively accessible, regardless of the internal state of the agent and perceiver. This emphasis on the phenomenology of the agent enables a clear connection between an individual's ethical values and decision-making processes and the milieu and the environmental consequences of a particular way of life. In other words, the concept of milieu rejects the abstract separation between processes that are often considered to be "internal" to the individual, such as ethical decision-making, and processes that appear to be "external" to the agent, such as environmental degradations and collective imprints. It enables to expand the realm of individual moral responsibility to tackle environmental harm caused by collective practices, as it will be shown in Chapter 4. If ethics emerges from our relation with the world, as Watsuji suggested, then artificially cutting off this relation to conduct decision-making as subjects on an object is likely to evacuate ethics from its very ground.

Moreover, this emphasis on the individual phenomenological experience anchors the concept of milieu at a local, concrete and particular scale. This focus on the milieu as local conceptually prevents homogenization and domination across communities and groups. Hierarchical and sometimes oppressive organizational patterns are highly likely to remain within groups, and other theoretical tools are needed to address them. Nevertheless, an analysis of the concept of milieu gives a safeguard against some worrying political and economic trends to intervene in far distant areas in the name of environmental protection. In contrast, every milieu is characterized by different systems of values and ways of life from which individuals can innovate and design locally and culturally appropriate more sustainable ways of life, and ask advice from independent scientific experts. The concept of milieu is thus *culturally sensitive*.

Third, the borders of milieus are porous, and milieus are closely interconnected in a *global network* continuously crossed by various flows (air, nutrients, goods, ideas, microorganisms, etc.). This is the spatial characteristic of the milieu. While milieus are local, they are constantly influenced and influencing distant milieus, as it clearly appears with global environmental changes such as climate change, and with the ubiquity of the Internet and communication media. Spreading through the multiple connections of this network and amplified by collective practices, the consequences of one individual imprint can be said to reach a global scale. This third characteristic insists on the interdependence of locally rooted milieus and distinguishes it from the concept of environment which, by virtue of already being an abstraction, can be uprooted into the idea of "global environment". Diverse milieus influence each other within an interconnected world. In this sense, the concept of milieu can be used as an *articulation between different scales of analysis and between different spheres of knowledge*. It synthesizes overlapping scales of patterns and processes: the scales of geology, climate, ecology and geomorphology; the scales of human settlements, land-use and sea-use and institutions; the scales of governance and regulation; and the scales of transports and communication (Harris 2007, 111). Each of these scales is usually handled by researchers and policy-makers in partial isolation, but all of them are combined in the local milieu.

Finally, a fourth characteristic of the milieu is its inherent *dynamic changes*. This is the temporal characteristic of the milieu. No milieu is fixed and permanently stable. Instead, milieus are changing through historical processes involving multiple agents across different scales. This contrasts with the idea of a pristine original nature to which we could – or could not – return. Indeed, the milieu is constantly changing, through the multiple interactions between its human and nonhuman inhabitants. In place of the picture of a linear evolution from a state of nature to a human-controlled world, the concept of milieu sets the stage for a constantly evolving and changing world in which multiple species adapt themselves, interact and shape their world (Imanishi 2013; Rupprecht et al. 2020). The dynamicity of the concept of milieu does not only fit the latest views of the world as imbricated nonlinear systems (Harris 2007), but it also justifies sustainable changes. Indeed, the milieu is not a fixed image grounded on local lifestyles and values from the past. Instead, it is an ever-changing web of meanings that are adapted by the currently living individuals to better fit their needs and values. If it is necessary and urgent to renounce to some seemingly old traditions and replace them with more sustainable practices, it should not be considered as a regrettable loss or a threat to a supposedly stable harmony, but as expressions of the ongoing cyclic relationships between living individuals and their long-lasting milieus. These considerations open the path for discussing sustainability and its articulation to ethics.

Table 2.1 Four characteristics of the milieu

---

1   Milieus are built through mutual relations.
    The concept recognizes the *interdependence* and *inseparability* of humans and their environment.
2   Milieus are experienced by phenomenological agents.
    The concept is culturally sensitive.
3   Milieus are interconnected globally in a dynamic network.
    The concept can be used as an articulation between different scales of analysis.
4   Milieus are continuously changing.
    The concept can be future-oriented and opened to *sustainable changes*.

These four characteristics hint at a paradigm shift from a worldview in which fixed and clearly defined things are what they are independently from their context, to a picture of an interconnected world with dynamic and changing flows interpenetrating and constituting elements, including ourselves. The first characteristic, namely that milieus are built in mutual cyclic relationship, shows that these processes are *constitutive* of the self and the milieu, which are concretely interdependent. The second characteristic's emphasis on the phenomenology of the individual sheds lights on our agency. We are not simply nexuses of various flows, but we can choose to close or open some channels, thus actively directing flows to shape the world and ourselves in different ways. These choices made by individual agents in their local milieus might appear insignificant, but they are connected in the large webs of interconnected and interdependent networks that span across the globe. These networks can be seen as interconnected channels that convey flows of goods and ideas with different resistance or fluidity. Finally, the whole system, by virtue of being composed of multiple diverse processes, is continuously changing. An in-depth discussion of this paradigm shift is beyond the scope of this book which focuses on environmental ethics. Still, these considerations place the framework of the milieu in the line of authors working on "process philosophy" (Winters 2017; Seibt 2020), and possibly also in the line of attempts to "replace values-centered ethics with a *dynamic flux ethics*" (Taylor 2015, 331).

The first, second and fourth characteristics oppose the idea of environment as a passive background to human activities. The emphasis on the phenomenological agent of the second characteristic rejects the idea that the milieu could be fully scientifically objectifiable, as it is sometimes assumed to be the case with the idea of environment. Crucially, the framework of the milieu does not reject the scientific enterprise that attempts to explain and model environmental phenomena and causal relationships as objectively as possible. On the contrary, scientific research is highly valuable and necessary to understand and assess environmental impacts, and the causal mechanisms that lead to what is judged to be environmental harm. Moreover, scientific expertise is also at the centre of designing more sustainable solutions. The emphasis on the phenomenological agent highlights the fact that *within human experience*, the milieu cannot be stripped of meanings and values. These meanings and values are attached to particular sociocultural contexts and worldviews that are not neutral and objective. If they were neutral and objective, decision-making would not be possible, as any decision is directed towards a normative direction, which is often left implicit. Recognizing that these normative directions, meanings and values are rooted in and constitutive of human experience urges us to unveil and closely consider these implicit normative assumptions for two reasons. On the one hand, these normative assumptions can provide us with guidance to ground environmental ethics. On the other hand, we must refrain from sweeping them under the carpet of apparently neutral and scientific assessments or we risk unilaterally imposing dominant normative values at the expense of underrepresented worldviews. Thus, the concept of milieu transparently encompasses the diversity of sociocultural and individual worldviews and values, in contrast with the scientific conception of the environment that tends to leave them implicit or even totally erase them.

Doing so, the updated concept of milieu echoes approaches that propose to integrate ecology and natural sciences, with humanities, social sciences and ethics, and developed concepts such as socio-ecological systems, landscapes and biocultural habitats. The idea of milieu partially overlaps with the concept of socio-ecological systems, that is, dynamic adaptive systems shaped by complex societal and ecological interactions (Colding and Barthel 2019). The socio-ecological approach has proven useful to study the sustainability of complex systems in terms of resource allocation, governance and relations between users (Ostrom 2015); biodiversity conservation (Hanspach et al. 2016); and protected areas management (Palomo et al. 2014). Yet, often, socio-ecological systems are studied with limited consideration for the internal state of the individuals (Manfredo et al. 2014). In contrast, as we have seen, the idea of milieu has at its heart the individual phenomenological perspective and agency. By addressing the phenomenological standpoints of the agents who experience and shape milieus, the framework of the milieu is complementary to socio-ecological approaches. Its emphasis on the phenomenology of the agent enables a clear connection between an individual's ethical values and decision-making, and the consequences of a particular way of life on the milieu.

Going one step further, the framework of the milieu can be applied in transdisciplinary research. For example, it can help articulate the standpoint of the individual agent with the complex dynamics of socio-ecological systems and contribute to identifying leverage points for sustainability in individual behaviours. It was already used to analyse concrete landscapes in Japan (Chakroun and Droz 2020). Within the framework of the milieu, the landscape refers to the concrete relations that link an area of land to the human and nonhuman agents, and reflects how the agents collectively relate to their milieu. In turn, experiencing and living in particular landscapes can influence behaviours and worldviews. Chakroun and Droz suggested that, through collectively and deliberately designing landscapes in a way that encourages individuals to adopt behaviours and worldviews beneficial for the maintenance and perpetuation of their milieu, landscapes could turn out to be highly strategic entry points to foster human-nature connectedness towards socio-ecological sustainability (Balázsi et al. 2019).

If we shift away from the phenomenological perspective of the human agent to approach milieus from the "observer" perspective, we can also approach them as biocultural habitats that host diverse multispecies inhabitants and that are formed through their interrelations. Biocultural habitats integrate, like milieus, a biophysical dimension, "from local ecosystems to the global biosphere"; a cultural and symbolic-linguistic dimension; and a socio-political, institutional and technical dimension (Rozzi et al. 2015; Rozzi 2018, 24). Like milieus, "any habitat influences and, in turn, is influenced by the ways in which it is inhabited" – that is, habits (Sala 2015, 239) – which are also influenced by "the co-inhabitants with whom we share habitats" (Rozzi 2018, 32). Rozzi proposed a biocultural ethics based on a socio-ecological approach of cohabitation within the biosphere and built around the concepts of habitats, habits and co-inhabitants. The latter includes other-than-human living beings:

> Co-inhabitants share habitats that they co-structure through co-inhabitation relationships. They establish ecological relationships of complementarity and reciprocity that occur through exchanges of matter and energy. Therefore, the care and conservation of habitats is the condition of possibility for the existence and well-being of the diverse co-inhabitants.
>
> (Rozzi, 2018, 39)

This complements the first characteristic of the milieu, which understands individual human beings as bound together with human and multispecies communities and their surroundings through mutual constitutive relationships. Similarly, Callicott recalls that "from an ecological point of view, an organism has been shaped and modelled – in its size, form, viscera, sociology, and psychology – by its interaction with other species" (Callicott 1997, 84). The inclusive view of the framework of the milieu accounts for these interspecies influences, as individual human beings are active parts of the multispecies community that inhabits a particular milieu. Regardless of the particular status given to other-than-human living beings, one cannot exclude them from the milieu. Milieus include other-than-human living beings (such as animals and plants), ecosystems, ideas of nature (such as conceptualizations of "Mother Earth" or materialistic view of the environment) and non-living and intangible realities like climate, which have impacts on living organisms.

2.3.3 Multispecies communities, ecosystems, animals and plants

The idea of milieu aims to transparently accommodate the diversity of world-views regarding human beings and nature, and to avoid participating in the dynamics of domination underlying conceptions that are claimed to be exclusive or universal. This concern converges to the ecofeminist literature. Ecofeminism foremost offers sharp criticisms of theories and implicit normative assumptions regarding conceptions of the relation between humans and the environment. Drawing on the feminist research on domination mechanisms, it tends to call for a more egalitarian and collaborative society in which there is no one dominant group (Merchant 2005). In general, ecofeminism tends to reject the sharp separation and dualism between human and nature. This dualism is often considered to be the source of a wide range of contemporary global problems, with global environmental problems at the front row. For example, Gaard and Gruen identified four problematic factors that contribute to causing the global environmental crisis (Gaard 2011). The first factor is the mechanistic materialist model that is supported by the scientific objectivation of the environment and that reduces natures to be "dead, inert" and turns its "oppression" into "a judicious use of resources" (Gaard and Gruen, 278). The second factor is patriarchal religions that tend to justify hierarchical domination of both women and nature as "divinely commanded". The third factor encompasses dualisms in general, including mind-body, female-male and human natures, because they often entail an ethics based on domination and control of one onto the other. The fourth factor is capitalism, because its normative direction of wealth creation is used to justify exploitation of people and living beings.

The conceptual framework of the milieu and the three-level model supporting it can be seen as a constructive proposal to go beyond the criticism of dualism while being aware of the dangers of implicit mechanisms of domination. The framework of the milieu makes way for concerns for domination and its potentially harmful consequences at different scales ranging from the self-identity to the relation between humanity and nature. Concretely, this concern is recurrent in my discussions of sustainability and responsibility in the following chapters. On this point, I go beyond Watsuji and Berque's discussions that tend to either brush the question away, or treat it in a way that could be interpreted to be highly worrying from an (eco-)feminist perspective. Taking seriously the global context of pluralism of worldview, the conception of sustainability in terms of milieu that I defend in Chapter 3 advocates a less intrusive and more pragmatic approach than ecofeminist critical theory that rejects bluntly capitalism and patriarchal religions, in order to be potentially acceptable for people whose belief systems include capitalism, patriarchal social organization or religions supporting hierarchical domination. Doing so, I do not reject the legitimacy of the ecofeminist argument *within some particular milieus*; I simply urge for precautionary stance at the global level, in order to avoid reproducing mechanisms of domination by imposing an anti–capitalist anti–patriarchal worldview on other cultures and milieus.

Beside ecofeminism, another important movement in environmental ethics that rejects a mechanistic and dualist conception of nature is Deep Ecology, which brings us to the debate between anthropocentrism and ecocentrism. Deep Ecology claims that living beings have inherent worth regardless of their instrumental value compared to human needs (Drengson et al. 1995). Various formulations of this claim became the central demand of ecocentrism, which built itself in opposition to anthropocentrism often understood as the claim that human beings are superior, or at least should be given priority over other species. In the introduction, we observed that values and meanings are not facts existing outside of our thoughts, but only beliefs that depend on our experience. It follows that values and meanings require thinking agents, namely, individual phenomenological agents. Thus, the claim that "living beings have intrinsic values", which is often held to be central to ecocentric theories, is deemed to be nothing but a normative assumption depending exclusively on the beliefs of the individual human beings defending this claim. In other words, natural elements can have "intrinsic" values only insofar as some individual human beings believe so. Thus, it *cannot* be argued for exclusively on scientific or factual grounds, but must be defended transparently as a normative claim held by human individuals and groups.

As a self-designed ecocentric theory, Deep Ecology rejects any claim of superiority of human species over other species in the natural world. The central tenet of Deep Ecology is that human beings are dependent on the natural world and part of it, as well as other living organisms. The founder of Deep Ecology, Arne Naess, wrote: "I am protecting the rainforest. [...] I am part of the rainforest protecting myself. I am that part of the rainforest recently emerged into thinking" (Seed and Macy 2007, 36). The realization of the intimate dependency to the environment would lead the individual to adopt a strong ecocentric perspective, and to "expand the self" in an identification process with "others", including other human beings, other animals, other living beings, ecosystems and, finally, the whole biosphere (Bragg 1996). Naess places the "relation" as essential for an entity to be itself.

> Organisms are knots in the biospherical net or field of intrinsic relations. An intrinsic relation between two things A and B is such that the relation belongs to the definitions or basic constitutions of A and B, so that without the relation, A and B are no longer the same things.
>
> (Naess 1973, 95)

It is impossible for A to exist as such, independently from its relation with B. Human beings are no more agents acting on a passive natural environment, but are an integrated part of it. Biosphere egalitarianism is another fundamental principle of Deep Ecology: Every living being has an equal right to live and flourish. Like most ecofeminist thinkers (Warren 1990; Plumwood

2002), Arne Naess is directly criticizing the conception of human, and self rooted in contemporary modern capitalist society and originated in patriarchal religious tradition placing human – and especially man – as a ruler and a master above the rest of the living beings. This conception of "ecological self" brushes past essentialism and embraces a spiritual-ethical tendency. The Deep Ecology agenda is thus openly different from mine, as it does not attempt to be compatible with the global pluralism of worldviews, or to build consensual common grounds between people holding different worldviews. Instead, it takes an interventionist stance as it implicitly suggests to replace worldviews and values of diverse human agents by their own, assuming that it would be a "better" world, an assumption itself resting on normative beliefs regarding what a better world consists of. In short, Deep Ecology does not seek to be sensitive or compatible with the global pluralism of sociocultural worldviews.

Two other criticisms that can be addressed in particular to the extended self of Deep Ecology are relevant for the objectives of my book. The first criticism is that if the self is "dissolved" in the environment, this may lead to a disappearance of agency, and thus responsibility. This relates to the debates around free-will, agency and ethical decision-making. The second critic has been addressed by the ecofeminist Val Plumwood under the term of *indistinguishability account*. According to her, Deep Ecology correctly recognizes the wrongness of the dualism separating human from nature, but its answer is a process of unification; "a metaphysics that insists that everything is really part of and indistinguishable from everything else" (Plumwood 1991, 13). By doing so, Deep Ecology is falling into a kind of atomism ignoring differences which are, in fact, a fundamental part of everyday life. This process of unification leads to a critical ignorance of distinctive particular needs. Jean Grimshaw presents it in a very concise way:

> It is important not merely because certain forms of symbiosis or "connection" with others can lead to damaging failures of personal development, but because care for others, understanding for them, are only possible if one can adequately distinguish oneself *from* others. If I see myself as "indistinct" from you, or you as not having your own being that is not merged with mine, then I cannot preserve a real sense of your well-being as opposed to mine. Care and understanding require the sort of distance that is needed in order not to see the other as a projection of self, or self as a continuation of the other.
>
> (Grimshaw 1986, 182–83)

Thus, we cannot simply identify ourselves with others and nonhuman living things and expect that our understanding of them through this identification process will be sufficient to adequately assess and address their needs. On the contrary, to avoid projecting our own situated interpretation on others, we need to distinguish them from ourselves. This objectivation process lies at the centre of the scientific methodological paradigm. Sciences can explore the particular needs of each organism while minimizing the bias of subjective sociocultural lenses. But sciences are still conducted by humans, and there remains an irreducible epistemological gap. Moreover, the process of identification with others does not go without risks, even between human beings. A similar criticism was addressed to Watsuji's conception of the relation between the self and the "totality" of the social group (Droz 2018b). The danger of erasing differences is that the dominant side of the relation might end up controlling the vulnerable other. Throughout her life, an individual concretely relational and closely intertwined with constitutive webs of relations with the others and the milieu is continuously vulnerable to changes in these relations.

More generally, what ecocentric and biocentric theories have in common is the tendency to value and the desire to protect nonhuman forms of life, such as animals and plants. This brings us back to the question of what is the place of animals, plants and ecosystems within the framework of the milieu. Nonhuman living beings are active parts of the milieu. Concretely, milieus are built through mutual relations not only between human beings, but between human beings and other living beings sharing the same spatiotemporal environment. The first characteristic of the milieu is the recognition of the interdependence and inseparability of humans and their environment, and that includes nonhuman living being, be it other species living together with us such as crows and chickens, or components of our bodily health, such as food and gastrointestinal bacteria.

Nonhuman living things also take active part in the dynamic networks connecting milieus across the globe. For example, migratory birds travel long distances seasonally and jellyfish can cross oceans carried away by streams, bringing with them other species and bacteria and influencing various milieus. Like humans, nonhuman living things are shaped by their environment and are shaping it by their usages, consuming food, building shelters and leaving faeces. They are also impacted and impacting the continuous environmental changes. Nonhuman living things thus play an active role in shaping the physical aspect of milieus lived by human beings. But they do not *deliberately* take part in shaping values and meanings shared by human communities. Interacting with animals and other nonhuman natural elements influences the worldview and imaginary of a human individual. Then, nonhuman living things can become central parts of the cultural imaginary and ways of life of sociocultural groups. Domestic animals and crops illustrate of how nonhuman living things are closely intertwined with meanings, values and worldviews of sociocultural groups.

Still, they are not phenomenological agents capable of ethical decision-making. They do not have access to our human discussions on meanings and values, and if they do have a consciousness and experience, their ways of "thinking" are likely to differ from ours. The second characteristic of the milieu focuses on the experience of phenomenological agents, that is, on human experience and ethical decision-making. Notably, the human species is an animal species, that is, human experience, by virtue of being irreducibly embodied, is possibly one particular kind of animal experience. Two questions appear: First, as human experience itself is highly diverse, what is to be considered the standard human experience? Addressing this question is beyond the scope of this book. For our purpose here, what seems to be central is the capacity to think, make decisions and take actions accordingly. From this perspective, animals and plants can be understood as agents leaving traces on the milieu, but they are not human agents capable of ethical reflection and decision-making, at least not in the ways we, humans, do. This provides an element of answer to the second question: Why focus on human agents? Other elements of answers are pragmatic. I am writing in a particular language, using modes of rationality and expressions that are human, to communicate, exchange and potentially convince other human beings. Plus, the goal of the book being to build a framework that could be globally consensual and lead to more sustainable ways of life, the audience to be convinced is exclusively composed of human beings.

Human beings also happen to be the main perpetrators of global environmental problems. Environmental "problems" are called "problems" because they directly threaten and harm human ways of life and projects. Simultaneously, our current dominant ways of life are one of the main causes of environmental problems. These are influenced by worldviews and ethical beliefs. Therefore, any long-term solution must involve ethics as a compass to direct people's ways of life. In other words, environmental problems are human problems, insofar as they are triggered by human actions, they are threatening things that matter to human beings and any solution to them must involve human agents. Yet, they are not problems of isolated individuals. They are the problems of human beings as beings who live intertwined in webs of complex and constitutive relations with each other, with other species and with the environment, in particular sociocultural and historical milieus.

The framework of the milieu is thus unapologetically anthropocentric, in the sense that, particularly in our daily life, we cannot think, perceive and act on the world from any other standpoint than our human embodied and socioculturally situated standpoint (Marietta and Maps 1995). We are always irreducibly trapped in our specific standpoint, and we frame our perception and reality in a particular sociocultural worldview. Of course, we can minimize the weight of the subjective and cultural lenses by using tools and seek towards objectivity, that is the core of the scientific enterprise. But when we understand, express and do something, we can never totally erase our standpoint, as it is inseparable from the fact of consciousness and thinking. Therefore, we are in a state of unavoidable anthropocentrism, because we *cannot* perceive, understand and act in the world without being human. Without a recognition of this fact, the normative attempts to shift to "ecocentrism" risk to impose one's lenses on others, be it human lenses on other species, or particular cultural lenses on other people's. As we cannot erase the human and sociocultural viewpoints, any claim to do so is likely imposing its own viewpoints without seeing it, as one's lenses tend to be invisible to their beholder.

This perspective does not imply that human must be the only holders of moral standings, or that values of nature must be derived exclusively in terms of how they can be used to fulfil human needs. From the perspective of the milieu, how humans and nonhuman things are valued is recognized to be dependent upon the milieu, that is, upon the webs of meanings and values that are situated at particular geographic and historical points. Values and meanings are not to be found, but they are to be assigned, and who is assigning them happen to be humans. One reason for this is that values and meanings are mediated by languages and means of expressions designed to communicate between human beings. I do not address directly the moral status of nonhuman living things in this book, but when we discuss the webs of meanings of our milieus, these *de facto* include meanings and values attached to nonhuman living things. Nonhuman living things are concrete parts of the milieus and indirectly approached and valued in different ways through the particular lenses of worldviews, in situated milieus.

These considerations urge us to recognize that we cannot escape wearing the subjective sociocultural human lenses for several reasons. For scientific and epistemological reasons, because thick glasses can blind us from some aspect of reality and hinder our knowledge of the world. For political reasons, because taking for granted a particular sociocultural set of glasses leads to the exclusion of others. For pragmatic and ethical reasons, because acknowledging our own situated standpoint is the first step in understanding others' standpoints and needs. Once we acknowledge that we see, think and act in the world through the situated lenses of a worldview, we can try to approach the bundles of intertwined facts and values that compose environmental ethics. The framework of the milieu understands these worldviews as being shaped through interactions between phenomenological agents, other species and elements, and milieus.

2.4 Medial matrix shaping individuals

The milieu as a matrix is nurturing human beings as persons. How the medial matrix shapes individual humans matters, because individual agents are taking ethical decisions and actions on the world. We need to understand how individuals are shaped and constrained by the medial matrix in order to unveil the scope of ethics. Cultures and social structures withhold some kind of agency only insofar individual agents are entrusting it to them. Before exploring how individual phenomenological agents act and contribute to creating the milieu, it is important to investigate how individuals themselves are products of their medial matrix. Here, I do not refer barely to the biological ways individual humans are dependent and made of elements of their physicochemical environment. Individual human existence depends on food, clean air and water, shelter and so on. Instead, I refer to the multiple socio-cultural influences shaping and modelling the individual as a person member of a particular group, the whole process being supported and mediated by a particular milieu.

This process can be examined from an observer perspective, as many social sciences, cognitive sciences, etc. seek to understand better how individuals are influenced by their environments. This can also be approached from the perspective of the phenomenological through introspection, self-analysis and narratives. My discussion here is based on works from social sciences, cognitive science and philosophy that are all largely informed by listening to and taking seriously the voices of the phenomenological agents. I articulate the exploration of how the medial matrix shapes the individual around three aspects: sense-making, the cultural imaginary and practices.

2.4.1 Sense-making

Any act of thinking seems to start with the question of how we make sense of the world, which includes ourselves. Sense-making also encompasses how we make sense of what and who we are. For a few hundreds of years, Cartesianism has been framing the question of thinking as a problem of an isolated consciousness. Yet, the well-known *cogito ergo sum* forgets the very fact that it is composed of words pertaining to a certain milieu and containing certain connotations and nuances. All these are everything but isolated. Even at a more basic level than language, what and how we are thinking is largely influenced by the circumstances in which we grew up and by our surroundings. Closing your eyes and focusing on the very fact that you are thinking, you may realize that you are still breathing and feeling your body, and you may be quite clearly aware of other people's presence or absence in your immediate surroundings.

The spectrum of ethics, namely how we should act and why, starts at the level of what and how we perceive and how we make sense of it. Answers to these questions will shape the scene and scenario in which we will make a choice and take action. As such, even sense-making has some ethical dimension. Only sense-making processes that involve at least a part of conscious thinking can have an ethical dimension, as ethics requires the possibility to choose differently, and the latter necessitates being conscious of choosing.

Authors such as Nick Chater and George Loewenstein suggested that our drive for sense-making is similar to our drives for air and food (Chater and Loewenstein 2016, 140–42). Most of the time, like digestion and respiration processes, sense-making is automatic. Yet, sometimes automatic information processing alone is insufficient. These times, we become aware of a gap in our understanding of a situation, and we are driven to fill it, spontaneously looking for the best-possible explanation in our memory or by redirecting attention. Multiple biases come into play, let alone the fact that making sense of something is generally pleasurable. This pleasure ranges from the ephemeral thrill of satisfaction when finding the correct answer to a puzzle, to the intense long-lasting pleasure coming from being able to give meaning to one's own life. The intensity of feelings of pleasure and pain coming from sense-making also depends on our expectations. Finally, as Chater and Loewenstein argue, we usually get more pain from a disappointed expectation (our inability to solve an easy mathematical problem) that we get pleasure from a successful sense-making far beyond our expectations.

Once we recognize our basic need for sense-making, it is easier to understand why it can sometimes be so challenging for us to face beliefs and facts that are going against some of the rock-bottom beliefs of our worldview, especially when it touches our worldview about our own life. When reasoning about a problem that does not have direct relation to our own life, a single new information can completely change our choice for a "best" theory. This is much harder when this new piece of information challenges the status quo of the explanation giving us maximum meaning to our life.

Encounters with challenging otherness are common. They can happen when reading a newspaper article, meeting someone who behaves in unexpected ways or even seeing an unprecedented natural event, such as a large pond suddenly drying due to the effects of climate change. As such, sense-making is always relational. We make sense of what happens, including of our own life, always in relation with others and other things, situated within a specific milieu. The context of any event, even of internal emotions, generally needs to be taken into consideration in order for us to establish our temporary best theory of the situation. This context is always medial. It can include other humans and other living beings, and it almost certainly includes others' ideas and environmental features such as landscapes, and more basic physicochemical elements such as the air. In other words, our awareness of our relation with our milieu starts at the level of sense-making.

If sense-making is relational, then what exactly is the participation of the other in it? Here, it is easier to start by observing what happens in our interactions with other humans. When we meet someone else, we are always situated in a particular milieu, and we are always embodied. Then, we tend to coordinate our behaviour with the others' so that the interaction goes by smoothly. This coordination necessitates both interactors to partially coordinate their sense-making. For example, as soon as a man comes to sit next to me in the train, I will slightly change my own position and adapt my behaviour to his presence following untold norms of politeness and mutual respect. These norms are strongly dependent on the particular milieu we are in, and also partly on the particular milieus we are coming from. Usually, all this process will happen without requiring any awareness of me. But if the man sits closer to me than what is dictated by these untold norms, then my own discomfort will make me aware of this silent violation of norms and I will consciously adapt my behaviour to it, moving aside.

In this situation, we are both making sense of the situation in our own ways. But at the same time, we are coordinating our sense-making by interpreting each other's behaviour. That is, "sense-making of the interactors acquires a coherence through their interaction and not just in their physical manifestation, but also in their significance" (De Jaegher and Di Paolo 2007, 497). Hanne De Jaegher and Paolo call this process "participatory sense-making". They argue that social interactions generate new domains of social sense-making on top of the one already available to each individual. The man sitting next to me and I are regulating our behaviours and sense-making around our respective intentions, the medial context in which we are situated and the social interaction itself. As De Jaegher writes, the social interaction "cannot be reduced to things like individual actors' communicative intentions" (Di Paolo and De Jaegher 2017, 89). In other words, my neighbour and I sustain the encounter by continuing to sit next to each other, and at the same time, the encounter is influencing each of us and invests us in the role of interactors. The social interaction appears as an "emergent autonomous organization" that is not reducible to individual behaviours.

We described before how we are in constant dynamic cycles of identification with our milieu, neither the relation nor the individual or the milieu having precedence to the other. Similarly, according to participatory sense-making, in social interactions, individuals emerge as interactors only within the interaction. Of course, each individual remains autonomous. But on top of their respective autonomy, the process of interaction gains a certain temporary autonomy. If one's autonomy had to be challenged for one to engage in a social interaction, then there could be no mutually respectful social interaction. The relational autonomy of this other is strongly diminished or even broken when she is seen as a tool or a problematic object for one's own individual cognition. Such a relationship has been diagnosed by various philosophers as unethical, such as Kantian argumentations of treating the other as an end and not merely a mean. But this point highlights the importance of the vulnerability in the encounter. Because of the intersubjective dynamic relation, there is always a primordial tension between the personal and the interpersonal domains.

If sense-making is a basic need, then the fact that we are taking a risk when encountering otherness that might bring us new challenging information comes as an evidence. The encounter with otherness can threaten our subjective status quo of best theory to make sense of a situation or of our life. Moreover, in cases where the other agent is reducing our autonomy, we may even be forced to reduce or change our cognitive engagement with the milieu. How we make sense of what happens is thus not a purely individual process, but it is participatory insofar as it also involves otherness and the process of social interaction. The participatory aspect of sense-making highlights the role played by common worldviews, norms and cultural imaginaries. These are dynamic processes supported by individual sense-making. Individuals also draw on some aspects of this common pot of meanings and explanations to form their own meanings.

2.4.2 Cultural imaginary

We all have a drive to make sense of events happening around us and of our own emotions and actions. Sometimes, those are pushing us away from our comfort zone to a place of uncertainty where our worldview is not sufficient to account for the strangeness of events. Then, we usually look for clues, explanations and answers in what I will refer to as the cultural imaginary. The cultural imaginary is the whole of imaginary representations particular to a sociocultural group. It includes myths, religious and spiritual beliefs and utopias, and generates common meanings to hold together the social life of the community. Individuals draw common references and meanings from the cultural imaginary that help them hold conversations, understand each other and develop a feeling of belonging to the community.

The cultural imaginary is best seen from the outside, in relation and in contrast to one's own imaginary resource and worldview. Paul Veyne uses the metaphor of the bowl inside of which we shape our thoughts and judgements, without noticing the very existence of the transparent walls surrounding us (Veyne 1971). Like goldfishes, we turn in circles inside our own sociocultural imaginary, and we watch in disbelief others' actions and reasoning, unaware of their own transparent fish bowls. Our own bowl is our basic frame of reference to understand, judge and act on our world. Compared to our personal worldview, the bowl of our cultural imaginary is wider, more ambiguous and inconsistent. Worldviews are a comprehensive conceptions of the world from a specific standing point. A worldview is a personal theory that we built through our own life to create meaning and guide our actions. It needs to be convincing and consistent to a certain point, at the very least to its holder. One's worldview is largely internal and personal, and can be accessed by others only through one's narrative and by inferences from one's behaviour.

On the contrary, the cultural imaginary is external and shared by a sociocultural group. As such, it is extrinsic sources of information. Clifford Geertz refers to it as cultural patterns that

> lie outside the boundaries of the individual organism as such in that intersubjective world of common understandings into which all human individuals are born, in which they pursue their separate careers, and which they leave persisting behind them after they die.
>
> (Geertz 1993, 92)

The cultural imaginary is historical, that is, constantly changing. It has been built by human intersubjectivity through history and appears through habits, norms, architecture, clothes and arts. The cultural imaginary is not only a floating cloud of stories, but it has a structuring function of the sociocultural reality that involves materiality. Institutions and buildings are inspired and informed by the cultural imaginary they are built in. Even though the cultural imaginary has been brought to us by historical processes far before our births, it is shared in the present by a community of living individuals and it carries the weight of the influences of past humans.

As frames of references for a particular community, cultural imaginaries are also structured around and changing with socio-political shifts. Benedict Anderson writes that, during last century, the frame of reference of "imagined communities" moved from the religious community and the dynastic realm to the nation. He argues that this shift was accompanied by three main ideas, namely, that truth can be accessed through science, that societies are moving through a homogenous empty time and that they are not "naturally" organized in centres (Anderson 1983). In other words, most cultural imaginaries nowadays accept some basic premises from the realm of sciences, like the fact that the world is a homogenous temporal and spatial space in which societies evolve and draw artificial borders around their claimed territories. Globalization is forcing cultural imaginaries to redefine themselves in relation to each other. There is simultaneously a broadening of the space of reference and an increase in contact with multiple different cultural imaginaries.

Here we must avoid the dangerous temptation of fixing borders to one cultural imaginary, as nationalism and populism do. Cultural imaginaries are "nebulas of disparate figures and historical phenomena" (Bayart 1996, 226), "magmas" of social significations that are both imaginary and meaningful (Castoriadis 1975). They are undefined, plural, multi-scaled and heterogeneous. Yet, what do they contain? Beyond moral fables and images of mythical creatures, they include ideal types of value systems and linguistic connotations. Language is a crucial element of cultural imaginaries, but they can neither be reduced to it nor be defined by it.

The cultural imaginaries in which we are brought up influence us beyond giving us role models, heroic stories and a precious mother tongue. Emotional and ideal connotations that words carry in addition to their literal meaning display such deeper meaningful complexity. Moreover, the cultural imaginary can even shape how we organize our perceptions and even perceive specific aspects by pushing our attention towards them or, on the contrary, leaving them in a shady ambiguity. The difference between the colours blue and green exemplifies such a case. The Latin languages make a clear distinction between these two colours, the blue of a clear sky and the green of grassy meadows. But in Japanese, this distinction is blurred and quite recent. The word for green *"midori"* is traditionally considered an exceptional nuance of the more common word for blue *"ao"*. Literally translated, apples, rice fields and spinach are all blue. Such ambiguous distinctions and connotations are included in the cultural imaginary.

What is the difference between the cultural imaginary and the milieu? Both seem to share a similar ambivalence. Cultural imaginaries are stuck in between materiality and imagination, in between being the cause and the effect of social phenomena and practices. Cultural imaginaries are flexible, fluid and creative flows of communal sense-making. They are the links between our imaginative drive for sense-making and the material reality. They are also the artificial decor for the play of the sociocultural narratives. Nevertheless, unlike milieus, imaginaries are by definition not present to the senses. They can influence perception and orientate attention, but they cannot create new external objects from scratch. They can be the motivation to do so, like in architecture, sculpture, fashion, etc. But in order to give birth to these external objects, the phenomenological agent must take elements from her milieu and transform them.

To make sense of their world, themselves and their milieu, phenomenological agents pick up storylines and ideas from the common pot of the cultural imaginaries they have access to. They choose them, adapt them and work on them to develop their own worldview. The individual experience remains subjective, but it is tinged by cultural imaginaries, structured by participatory sense-making and situated in a particular milieu. The web of memories carried by the phenomenological agent is unique and personal, but the memories and the way of remembering them are influenced by the cultural imaginary in which the self is immersed. When acting in the present, the phenomenological agent is weaving and orienting her future by reflecting on and reinterpreting cultural expectations. Cultural imaginaries are one of the ways in which the milieu is historical; another is the temporal existence of the material world. Individuals learn from the meaning, values and usages already existing; adapt these to present situations; and change the material reality. Another crucial difference between milieus and cultural imaginaries is that the latter can be displaced while the former is anchored in a specific geographical location. When communities and sociocultural groups are migrating, they bring along their own cultural imaginaries into the new milieu where they settle down. Diverse cultural imaginaries can cohabit, exchange influences and shape together a particular milieu. Cultural imaginaries are not only producers of meaning, but they are also formatting social practices.

2.4.3 Practices

Phenomenological agents are shaped by the medial matrix insofar that they make sense of the world around them and of themselves through dynamic processes that are participatory and situated in a specific sociocultural and environmental context. To do so, they also borrow conceptual tools, images and guidelines to the cultural imaginaries they can access. They are also agents, and this very fact is irreducible to who they are. As long as they are alive, individuals are constantly acting. Their actions are always taking place in a particular milieu, involving social interpretations and material goods.

For some important decisions and actions, we consciously balance different options, pick one up and act on the world while being aware of the process, the action and the probable consequences. But most of the time, we act automatically. We may come to the table, pour a glass of water, take chopsticks and start eating while being absorbed in a conversation and without questioning our actions. We just follow the flow of habits. These automatisms often become apparent when the habits are broken by external circumstances. If you cannot find chopsticks, then you might wonder what other ways you have to eat your food properly. You might feel confused about how to use properly forks and knives, or your bare hands. In public, you might worry about what other people think of your manners to the point of refraining from eating despite feeling hungry. This sheds light to the wide extent to which social meanings are guiding our everyday life actions.

Repeated in the everyday life, these habits are practices. Practices are heavily charged with social meanings carried by the milieu. From the perspective of the phenomenological agent, they guide and constrain one's behaviour. By observing others, the agent can learn the table manners, adapt her behaviour and eat in a way considered proper. By integrating a specific practice of table manners, she also reproduces it and reinforces the fact that they are the "proper" way to eat. Conversely, the same table manners that she uses as guidance can become constraints. They might deter her from eating with her hands, even if she is hungry or if that is the way to eat it that gives her the most pleasure. Yet, simultaneously, she can be aware of the practice and make the decision to resist to it and contest it, and bite in the raw apple at the dinner table, enduring the social consequences in the form of disapprobatory looks and verbal reproaches. As such, actions of an individual in relation to the practices of the specific milieu she exists in are the site where she exercises her autonomy. The flip side of this is that individual actions are probed to normative judgements and to ethical justifications, while the same goes for practices themselves. Practices can be normatively challenged, contested, resisted and opposed by individual actions.

Individual actions and social practices are spanning on a spectrum ranging from completely deliberate behaviours explicitly and intentionally coordinated with the social norms, to automatic habits resulting from internalized social meanings. Sally Haslanger suggests that practices consist of both social meanings as guides for coordinated actions and the management of goods, their production and sharing (Haslanger 2018). These two aspects are held together by a looping effect that makes the practices self-sustaining and relatively stable. For example, if a social group considers chopsticks as essential tools to eat properly, then chopsticks will be produced and made available in the places where these individuals eat. Their availability will then reinforce the practice of using chopsticks.

The materiality of resources needed to sustain practices has important consequences. Resources are subject to scarcity, that is, a site of competition and distributive justice. Besides, their production and usages may have consequences on the natural environment. For example, the production of disposable chopsticks made from wood is a cause of deforestation and unsustainable management of natural resources in some areas. But this production is required by the consumption of disposable chopsticks, namely, the practice of using them once only and the social meanings attached to it.

Practices are a stage setting for facilitating coordinated actions of the agents. The practices and social meanings associated with them are locally transmitted and usually enable smooth social interactions. From the perspective of the observer, it might seem that all agents involved in the practice share the same intentions. Actually, different agents involved in a same practice may have different reasons for it. They might be confused about it and interpret it in different ways. Practices "structure the possibility space for agency" (Haslanger 2016, 127) but do not determine the individual's intentions and inner decision-making. They are toolkits of habits that the individual chooses to follow, to slightly change or to frontally resist. Like cultural imaginaries present to the individual a variety of possible images, explanations and ideals, practices offer her a whole range of guidelines for behaviours. On the stage set by practices, agents are choosing to impersonate the suggested roles or not, and cope with the social and material consequences of their decisions.

The intentions and reasons held by the phenomenological agent when engaging in a practice do not always reflect the social meanings that will be attributed to the behaviour. If I hug and kiss on the cheek my Japanese friend when I meet him in rural Japan, even if my intentions are to greet him amicably as people do in my home country, himself will be surprised and might feel aggressed, while bystanders may be shocked and openly disapprove my behaviour. An important ethical consequence of this fact is that the agent's intentions cannot be used as exclusive explanation and justification for her behaviour and for how her behaviour is perceived by others.

Furthermore, practices are not existing on their own, but intertwined in networks of practices located in a specific milieu. This network can be referred to as a social structure. As practices are behaviours conforming to some cultural schemas in close relation to material resources, social structures have important material relata and consequences. Social structures unfold in relations between people and in relations to things, especially to resources. That is, the social structure can contribute to injustice, even if the agents involved in the practices composing this structure have no intention at all to commit actions that have harmful consequences.

When it comes to environmental issues, instances where the social structures themselves have ethical negative consequences are more rule than exception. For instance, the practice of driving a private car every day is sometimes grounded on social meanings associating the ownership of a private car with adulthood and autonomy, and on the material resources making it possible, from the affordability of the cars to the availability of the roads. These social meanings and resources supported the emergence of a practice that, in turn, contributed to designing the landscape in a way facilitating the continuation and spread of this practice. This led to a critical situation in which individual agents may be aware and willing to reduce their greenhouse gas emissions, but cannot give up their private cars without moving to an urban area, because no public transportation is available.

In short, practices are self-sustaining, as they acquire stability by building social structures and schemas. The performativity of practices is what links the conceptual levels of sense-making and cultural imaginaries to the concrete consequences of common actions. The medial matrix is shaping how individuals make sense of the world and themselves, what they can imagine and how and why they behave. The three processes of sense-making, interactions with cultural imaginaries and practices are closely intertwined threads that form a web of significations and guidance for action strongly attached to a specific natural environment. This is the milieu as the complex and dynamic matrix from which we build our identity and manoeuvre throughout our life. The milieu is what appears in our subjective relationship to space and time, when we progress in life tangled up in these webs of incoherent significations and interpretations.

These observations about the power of networks of practices, the pervasiveness of cultural imaginaries and the participatory aspect of sense-making have important ethical implications. I previously suggested that ethical agency rests at the individual level of the decision-making and action of an embodied agent. But the fact that our actions can be interpreted in ways different from what we intended to do and the fact that they are importantly guided and limited by the social structure we are living in entail that moral responsibility cannot be reduced to the isolated actions at the individual level. We have also, as groups of individuals, a responsibility about what "we" can and should do together to improve the social structure and the milieu. Because, as Haslanger writes, "if we attempt to change how we perceive and think without changing the social reality that is responsible for the schemas we employ, our efforts are unlikely to be sustainable" (Haslanger 2015, 12). If it is not for common actions targeting the social structure, our individual good will is likely to wear out confronted to the self-sustaining loop between material resources and symbolic reality. These considerations bring us to the other side of the coin of the milieu, namely, how the individuals and groups of individuals are shaping the milieu by constantly challenging it and leaving the traces of their existence on it.

2.5 Ethical action as medial imprint

Any human individual takes action on the world, and these actions have consequences. The consequences of one action can touch the individual in her identity, social status and biological existence. They can also affect the medial level, namely, the sociocultural webs of significations that are attached to a specific milieu. Besides, they often also impact the natural environment. Most actions affect all these three levels, yet more or less directly and at different degrees. That allows me to refer to them under the umbrella term of medial imprints.

The phenomenological agent is rarely aware of all consequences of her own actions. This limited awareness has two main reasons. First, most of the time, the phenomenological agent acts automatically without reflecting or actively choosing to engage in a specific practice. Ingrained habits, skills and dispositions guide how individuals perceive and react to the social world around them, as captured by Bourdieu's notion of habitus (Bourdieu 1990, 53). Sometimes, the phenomenological agent deliberately engages in an action, weighing the pros and cons (e.g. to reach a reflective equilibrium (Rawls 2005)), by thinking about the possible consequences and making a conscious decision to act. Yet, even a fully well-intentioned agent situated in positive circumstances putting her in a position to judge to the best of her knowledge may not be aware of *all* the consequences of her best-possible action. The second reason for this state of affairs is that an action can have domino effects that are difficult to predict. This state of partially irreducible and partially voluntary ignorance is ubiquitous with consequences on the highly complex webs of relationships of the milieu and the environment.

Any ethical system addressing environmental ethics is frontally confronted to this partial disconnection between the intentions of the phenomenological agents and the medial and environmental consequences of her action. The assignment of ethical responsibility is complex, yet unescapable. Now, I will first discuss the types of medial imprints and the domino effect typical to most environmental issues. Then, I will explore the phenomenology of the ethical agent, searching for a way to bridge the gap between an agent's intentions and the consequences of her actions. Finally, I will develop implications of these discussions for ethical responsibility.

2.5.1 Domino effect of individual actions as medial imprints

Action is a bodily movement. In philosophy, it is often restrained to bodily movements that are conscious, intentional and subjectively meaningful. As consciousness and intentionality are vaguely defined and hardly clearly identifiable, I set aside these criteria and follow enactivism theories that understand perception as active, because it involves sensorimotor processes and it is goal-oriented. This understanding fits the pervasive and participatory aspect of sense-making and practices. In this widened definition, actions are bodily movements that *can be made* conscious and intentional by a shift of the agent's attention. When becoming conscious and intentional, they instantly become subjectively meaningful. As such, sighing and taking a deep breath are actions, while breathing during sleep is not.

Actions are essentially relational, building a relation between the agent and some otherness. This otherness can be the agent's own body, external objects, living beings and human individuals, etc. Because they involve the body, actions always have consequences. When exhaling, we reject an air richer in carbon dioxide than the air we breathed in. During our time awake, we act almost continuously, spreading innumerable multitudes of consequences.

Most of them are or at the very least seem to be benign and devoid of any significant effect – they seem not to trigger any change in the state of the world. Some have enormous consequences, like setting fire to a house or jumping off a cliff. Any action is subject to normative assessments, considering the intentions of the agent and the apparent consequences. Normative assessments can be conducted by the phenomenological agent herself and by other observers, prior, during and after the action. Prior assessments are conditional, but clearer about the intentions of the agent, while post-assessments are clearer about the consequences, but the intentions of the agent might be obscured by various cognitive biases.

Regardless of the intentions of the agent, many of her actions have significant medial imprints. Medial imprints are traces left on the milieu by actions of human agents. Here, I explore the simplest type of medial imprint, namely an action by an individual. Medial imprints of social structures and groups of individuals when no individual action can be singled out from the consequences will be discussed later. Of course, completely singling out an action from its context is depriving it from most of its meaning. Actions are closely dependent on needs, goals and resources of the agent, and on the tight web of causal relationships of the world. An easier approach is to watch relevant consequences from the perspective of the observer.

The material consequences of an individual's action can be distributed geographically. I distinguish three geographical scales where these consequences can be found. First, the local scale consists on the immediate surroundings of the place where the individual acted. Depending on actions, what is local can be a room, a village, a city or even a region. Second, an action can affect another local place that is distant from the place where the individual acted. That can be the neighbour village, a city in another time zone and a region of a rainforest on another continent. Third, consequences can be global, such as climate change. Then, to take back my three-level model, an individual action's consequences can affect mainly the environment, or the milieu, or the individual. The three levels – environmental-medial-individual – are co-dependent and inseparable as soon as there exists an individual agent. It follows that any intentional and meaningful action is likely to have consequences spanning on the individual and medial levels, and often on the environmental level, even if the effects at the latter might be insignificant.

Pollution and harm to ecosystems can be cases of local environmental impacts of an individual action. For example, a farmer in a mountainous village uses an herbicide in her field to cultivate corn. In this fictional example, she repeatedly does the action of spreading the product on her field. Her intentions are to increase the yield of her land and her monetary income from selling the corns to be able to afford the education of her child. Regardless of her intentions, she may or may not notice the soil degradation in her own field. Another local environmental consequence is the local extinction of a specific species of wild flower in the surroundings of her field (see Table 2.2). Her family may regret the disappearance of this flower, because it was emblematic of a season and of the village identity. This is a medial local consequence of the usage of herbicide in the field. Another medial local consequence is the loss of local knowledge related to weeding. Because it seems to be more convenient to use herbicide, the farmer may not feel the need to teach her children or her helpers how to use traditional tools and technics. If unstopped over a few years, the continuous usage of the herbicide may impact the physical health of children, and lead to soil biodiversity loss and to the death of the soil. Due to the soil infertility, the family may then suffer from poverty and hunger.

The usage of herbicide in the farmer's field also has consequences in distant localities. On the environmental level, it might pollute underground waters, affect distant ecosystems and induce biodiversity loss. On the medial level, the water pollution may affect other distant communities and force them to change practices or even to migrate. Finally, distant and local consequences might affect the individual directly in the form of stigmatization, social pressure and exclusion.

Consequences of an individual action at the global level are generally coming from domino effect. For example, the reduction of biodiversity and the drastic diminution of the population of insects in the area may hinder its usage as an important resting spot for a species of migratory birds. The migratory birds could be forced to change their migratory path, which might threaten the survival of the species. Other ecosystems around the Earth might be dependent on the seasonal passage of these migratory birds. It would be exaggerated to say that the precise individual action of spreading herbicide in one's private field is solely responsible for this whole domino effect, but it would be equally inaccurate to say that it does not play any role in the occurrence of this consequence. The same goes for global problems of desertification and habitat loss.

Table 2.2 Examples of consequences of an individual action

Usage of herbicide in field	Local	Distant + local	Global
Environmental	Soil degradation in the field	Pollution of underground waters affecting distant ecosystems	Changes in species migration paths Desertification
Medial	Loss of an emblematic wild flower, loss of local knowledge	Water pollution affecting other communities and inducing forced migration	X
Individual	Illnesses and inflammation, hunger	Stigmatization by other communities	X

Global medial consequences also affect milieus at the global level, but it is hard to imagine how the particular individual action of spreading herbicide could reach the global scale. Globalization offers many examples of effects on the milieus at the global level. Yet, they are not the consequence of one individual action. With the ubiquity and accessibility of social media and of buzzing news, ideas, images and sounds circulate between milieus rapidly at the global level. Yet, this circulation is the product of tens of thousands individual actions of "sharing" and "commenting" information, not of one single individual action. For the same reasons, global consequences of an individual action affecting the individual level are unlikely.

The example of the usage of herbicide sheds light on the inseparability of the three levels of analysis (individual, medial and environmental). When perceived, described and discussed by the community of the farmer, all these consequences constitute her individual medial imprint. Her individual medial imprint is the whole of the traces her actions leave on her milieu. Because the milieu is a shared place of intersubjectivity, her medial imprint also includes effects that she is not aware of. The local extinction of the symbolic wild flower could have a powerful impact on the local milieu, even if the farmer only partially, or not at all, understands the role she played in it.

For most environmental problems, domino effect across geographical scales is striking. Climate change and sea-level rise are such typical examples. Because the Earth ecosystems and milieus are all closely interconnected and interdependent, an individual action might seem to have enormous consequences on the other side of the planet. But it is crucial to set a limit to prevent interdependency to be used in a *reductio ad absurdum* argumentation that would forfeit responsibility of individual agents. The fluttering of a butterfly's wings does not set off a tornado on another continent on its own.

Contrary to butterflies – supposedly – humans are constantly observing each other, making sense of the world together and assessing normatively each other's practices. Technologies of communications have made this phenomenon of exchanges stronger than ever before. Today, the video of the fluttering of a butterfly, shared by the clicking actions of thousands of individuals and promoted by the algorithms of social media companies, might take some platforms by storm. It might become a part of a cultural imaginary or echo in the imagination of an artist who might produce a sculpture that is to become a famous part of a culture's tangible heritage. But it is clear that all of these are *not* the individual medial imprint of the videographer. The videographer only took a video and posted it online. It was then exchanged and shaped by thousands of others' actions (clicking, commenting, painting, writing, butterfly conservation campaigning...). Still, without the action of the videographer, it is unlikely that the sculptor would have created this exact work at this particular time. I refer to these domino-effect consequences mediated by other humans' actions as the mediated medial imprints of the videographer as a member of a social structure and a milieu.

In sum, every individual action is likely to have a medial imprint. Throughout one's life, the individual does a collection of actions guided by practices and inspired by elements from the cultural imaginaries intertwined with participatory sense-making. The traces left by this whole set of behaviours of one individual constitute the individual medial imprint.

2.5.2 Inaction and mediated imprint

I distinguished two types of medial imprints. First, individual medial imprints are all the significant traces left on the environment and on the milieu by the action of an individual, and which do not involve other humans' actions after the individual's action. I consider as significant any effect that is important enough to cause irreversible damages (such as a species extinction) or to trigger domino effects resulting in changes of the milieu or of the environment affecting other species' survival and other humans' health. Some actions have domino-effect consequences spanning beyond the local scale of the action until the global level. According to this definition, breathing generally does not produce medial imprints.

Second, most individual actions and behaviours have mediated medial imprints. Mediated medial imprints are the significant impacts that an individual agent has by virtue of being a member of a social structure. The social structure is a network of intertwined and mutually supporting practices, most of the time located in a specific milieu. Because of this complex intertwinement, an individual does not need to do any action to have an impact on the world. Because of the close interconnectivity of our social structures, the very existence of an individual influences the social structure and the milieu. A quick thought regarding how innovation and production of goods are designed after market analysis of consumer demands makes this blatant. Regardless of our intentions and actions, as long as we are alive, we are considered by others as potential players in the milieu. Others may try to influence our behaviours or take advantage of our inaction, and in any case, they take us into account in their own decision-making. This does not mean that others have negative intentions towards us, they actually might ignore us deliberately. But we are sharing the same space of existence in the form of globally interconnected milieus, so we expect others to conform to some basic norms of behaviours.

In the case of market analysis for launching the production of particular goods, the individual is merely a number in a calculation table. Her intentions are predicted and modelled to be potentially manipulated by marketing strategies. One does not need to do anything to be counted as a potential consumer. Then, the number of potential consumers will be used as a basis to calibrate production, even if the individual later never buys the goods. Still, these useless goods are produced, using raw materials, energy and workforce, and necessitate proper disposal facilities. Public opinion and reputation are also ways in which the bare existence of an individual influences others' behaviours. Simply standing next to the river and watching it flowing might deter a bystander from throwing plastic waste in it, by fear of being judged. The simple fact of being in a specific place influences others. Depending on the appearance of your body and clothes and on your social role, the way others adapt their behaviours will change. What can be done to an individual by others, what can be done in her name and what she actually can do all partially depend on her social role and position in the society.

Not doing something has also consequences. Not helping a child drowning when having the possibility to save her is consensually wrong. Conversely, refraining from using plastic bags and refraining from owning a private car in rural France also have medial impacts. This is not only because one plastic bag is spared or because there is one less car on the roads, but more importantly because other people observe the agent's behaviour, assess it, judge it, etc. The presence of an individual who does not comply with certain particular practices in a milieu triggers reactions by other individuals sharing the milieu, ranging from passively noticing the difference to anger or re-evaluation of the merits of the practice in relation to one's own behaviour. Not complying with an expected practice can have mediated medial imprints.

We could argue that the consequences of the action of the farmer of spreading herbicide in her field are not only her medial imprint, but also the mediated medial imprint of the designers of the product, the salespeople, etc. Furthermore, we could argue that they are the consequences of the whole global social structure imposing conditions of poverty, consumer demand for corns and costly expectations for children's education. Giving the same weight to medial imprints and to mediated medial imprints risks to face the problem of *reductio ad absurdum*. In the normative assessment of one's behaviour, including one's own, medial imprints are generally given more importance than mediated medial imprints. This is understandable because it is easier to assess direct medial imprints than mediated medial imprints.

If we consider normatively the consequences mediated by an individual's existence, then the problem of where to draw the line is flagrant. A well-known saying reads: "The world is a dangerous place, not because of those who do evil, but because of those who look on and do nothing". My discussion of mediated medial imprints converges in this direction. Are we all partially responsible and normatively blameable for the consequences of the farmer spreading herbicide on her private field? How responsibility should be distributed? Up to now, we have been investigating the individual medial imprint from the perspective of the observer, by listing the consequences of one's actions, inaction and existence. The perspective of the phenomenological agent might bring some further elements to sketch the line of what counts in the normative assessment of individual behaviour.

2.5.3 Phenomenology of the ethical action

The phenomenological agent is always embodied and situated in a specific milieu that contains practices as guide and constraints and cultural imaginaries as shared frames of references. She makes sense of her world in relation with others, through dynamic processes in which social interactions play a key role. Yet, the range of what the agent is aware of at a given time is much more limited. The spotlight of our attention makes visible and consciously accessible to us only a very small part of all these intertwined signifying relationships that make the milieu. The metaphor of the spotlight is used to highlight the contrast between "endogenous focus and exogenous openness, self-conscious planning and unself-conscious absorption, spotlight and lantern" (Gopnik 2011, 132). As Alison Gopnik writes: "if we change the way we think, we also change the way thinking feels to us" (163). What enters the window of awareness of the agent is often partially determined by what are her goals at this specific time. The environmental surroundings are always present to the attention of the agent, but they appear covered with affordances. The process of sense-making is hardly noticeable, except on the rare occasions when the agent thinks about her own thinking. The cultural imaginary, like a fish bowl, is usually invisible to the agent. Most of the time, practices are executed automatically, without entering into the spotlight of the agent's attention. Most of the agent's actions are done in an automatic manner, without deliberations.

The spectrum described in the case of practices applies to the phenomenology of the agent. An agent's actions range from unaware automatic actions resulting from internalized practices, to conscious and deliberate actions that the agent explicitly does in relation with different particular elements of her milieu. In the case of automatic actions, the agent expects with high certainty an outcome. If the expected outcome does not occur, the agent may shift her attention to the action. For example, if no water comes after turning the tap (a usually automatic action), the agent will shift her attention from her ongoing thoughts to the action of turning the tap. Exceptional inconsistencies with the expected outcomes of a habit frequently move an action from the automatic grey part of the automatic-deliberate spectrum to the centre of the spotlight of deliberate action.

Most of the time, the phenomenological agent is unaware of most of the consequences of her action, even less of its domino-effect consequences. The effect that she is usually aware of, or that she can easily shift her attention to, is the intended outcome. The farmer is likely to think only about getting rid of weeds while spreading the herbicide (see Table 2.2). Other consequences might never come to her mind. Even when seeing them happening, she might not make the link with her action of spreading herbicide. She can become aware of some of other local consequences, such as the probable disappearance of a wild flower from her field. But most of the consequences are far beyond the limits of her knowledge. For example, if she does not know about the existence of underground waters in her area, she cannot predict the impacts of polluting them on distant ecosystems and on other human communities dependent on them. Needless to say, global environmental effects such as changes in species migration paths and desertification are unlikely to appear in her range of knowledge unless she specifically researches about them or she previously received education about them.

Thus, the agent is largely blind when taking most of her actions. Her blindness can come from automatism and lack of knowledge, but it can also come from the transparent walls of the bowl of her cultural imaginary. An old man who spends his life in the countryside surrounded by stories, images and other people giving utmost importance to the ownership of a private car on the path to success, its need being confirmed by the state of affairs of urban sprawling and lack of public transportation, might not be able to envision a fulfilling life without owning a private car. He might not be able to imagine leading his life without it. Dogma can also blind the agent by hindering her from imagining a possibility hidden by a rock-bottom belief. For instance, the rock-bottom belief in economic growth as the unique relevant measure of happiness and goodness can shade away contradicting facts, and one's own confirmation bias might push the believer to actively seek and favour any interpretation that might support her rock-bottom belief.

On top of that, what is accessible for an agent at the moment of doing an action is often determined by the social role she is endorsing at this precise moment. The agent might have the required knowledge about a question when she endorses another social role but lose awareness of this knowledge when shifting roles. For example, a government official may not see the urgency of dealing with some safety issues in nuclear power plants, because she is embroiled in other deadlines and tasks, and her role is, in her own understanding, clearly delimited to apply rules and follow orders, not to bring to the attention of her superior such a file, and even less to treat it despite contrary orders of her superiors. When coming home to her children and discussing with other parents, she might agree on the urgency of governmental action on nuclear power plants' safety. But each opinion being strongly linked to the social role endorsed when holding it, it might never occur to her to address her worries as mother when she is a governmental official. Certain questions are associated with some specific social roles, and hardly pierce through social roles.

In short, the process of thinking of the agent is situated in a specific cultural imaginary, structured by a specific worldview and grounded on some rock-bottom beliefs. The perspective taken on this stage is influenced by the intended goal, by the social role endorsed at the moment and by the resources accessible by the individual at this moment, ranging from ideas to affordances. The length of the deliberation process is also limited by circumstances that can be external (deadlines, weather, etc.), bodily (tiredness, hunger, etc.) or self-imposed. Finally, the conceptual tools accessible to the agents are largely shaped by all of the above and more general elements such as language, imagination and knowledge.

No individual situated thinking process is covering all of human knowledge and the conditional possibilities of consequences. No individual thinking process is fully coherent. So the rationalist requirement for coherence in one's reasoning needs to be nuanced, because it is unrealistic (Bourdieu 1977; Fournier 2012). If rational coherence is practically impossible to reach, it can still be an ideal to aim for. But it is crucial to keep in mind the fact that even if the deliberation and the final best-possible decision seem to be fully rational and coherent to the individual, it is unlikely to be the case. Indeed, as discussed in the introduction, any agent is thinking in terms of her own system of rationality. Moreover, even if a specific reasoning seems to be fully coherent and rational to a group of agents, it might not be the case to another group. Caution thus needs to be taken when requiring coherence and rationality from some agents in order to avoid simply imposing on the agents one's own theory of rationality and associated rock-bottom beliefs and values.

What role do ethics and morality play in this picture? Besides unexpected outcomes, there are other ways in which an action can move on the automatic-deliberate spectrum to the centre of the spotlight: social pressure, moral blame, and moral shock. If some members of the milieu disapprove an action, we are likely to deliberate about the merits of doing it, balancing it with the social and emotional losses of blame, shame and loss of trust and complicity. Any sudden change in practices comes with social consequences in the local milieu, as it can be perceived as a challenge to accepted social norms. What counts as a protest for different people covers a wide range of actions. Social risks associated with protest are particularly high when the cultural imaginary includes only few stories of successful resistance bringing positive results for the community and a shortage of common stories incorporating a similar ethical justification to the one the agent. Disobeying takes enormous amounts of energy and can be very subjective (Loizidou 2013). These risks are not to be underestimated, as individuals engaged in public protest regarding environmental issues are regularly exposed to serious threats to their life and to their relatives' (Larsen et al. 2020).

Despite these risks, moral shock can trigger action and change in practices. James M. Jasper uses the term of moral shock to refer to the cognitive and emotional process encouraging individuals to actively participate in social movements (Jasper 1997). He argues that moral shock often forces individual to articulate their moral intuitions and to reassess some of their behaviours and inaction, up to the point of engaging in moral protest. For example, watching a documentary about how to deal with nuclear wastes and to communicate their dangerousness to humans in the far future may trigger moral reconsiderations and self-reflection, up to the point of reconsidering one's own lifestyle, electricity consumption, voting habits, etc. Moral shocks can be turning points in the individual's biography (Droz 2018a), and starting points from where the agent develops strategies to fulfil the sudden drive for ethical actions, considering the available financial, educational, technical and time-related resources. To be able to take an unusual action challenging her habits, the agent needs time to assess the situation, to clarify her goals, to identify the material, conceptual and social resources she has access to, to predict others' reactions to her change of habit and to decide, over and over again, if she goes on with her action or if she gives up.

The phenomenology of the agent reveals chaotic and confused decision-making processes for most actions. The agent is partially, but irreducibly, blind to the possible consequences of her actions, and she is also emotionally and socially deterred from going out of the comfort zone of her culturally sanctioned habits. Moreover, the phenomenological agent develops her identity through her choices of behaviours. Changing a habit might require to question one's own identity. As human selves are in constant dynamic cycles of codetermination with the milieu, agents constantly edit their identities to adapt to changes in their milieus and to changes in their subjective understanding of the world. Yet, most of the adaptation processes happen automatically or semi-automatically, supported by the social structure. Deliberately swimming against the tide of the social structure we depend on can be emotionally exhausting and socially dangerous.

Conversely, overcoming this tremor can be fulfilling. On the one hand, temporarily and partially switching to the observer to assess one's own behaviours is overwhelming. On the other hand, the individual realizes herself by acting on her milieu. When deliberately pulling a thread of the ecological, social and historical web of relationships that is the milieu, the agent becomes aware of the interconnectivity of the world and of her own agency. The comprehension that the agent does have a lasting imprint on the milieu can be a very meaningful experience. The individual is always embodied and situated spatially and temporally, but both the past and the future are deployed in the interconnection of actions. Watsuji even referred to this as the "structural moment of human existence" (Watsuji 2004, 1). The milieu is something by which the individual agent is objectified. It is also that by which the individual agent's existence is spanning beyond her spatiotemporal point of observation, towards distant places and distant future, through the whole of her medial imprints and mediated medial imprints.

2.6 Summary

I develop the "three-level model" based on three densely interconnected levels, like concentric circles around the human individual (1), always surrounded by the milieu (2) and situated at a point in the environment (3). The individual human is key in the three-level model, because it is the place of ethical agency, that is, perception, decision-making and action. Depending on which perspective we take, be it the observer perspective or the phenomenological agent perspective, questions and situations take very different colours. I also insist on a non-atomistic and dynamic conception of the self that is intrinsically relational. Indeed, we are always anchored in one body situated in a specific spatiotemporal point in the world (namely, the milieu and the environment), in between perception and action. To make sense of the succession of events of consciousness, memories and future aspirations, we are constructing our idea of ourselves that we also use to normatively justify our actions.

The Japanese philosopher Watsuji Tetsurō (1889–1960) developed the concept of *fūdo* (風土), that I translate by "milieu" following Augustin Berque's work. The milieu is the "environment" as it appears covered by web of significations and symbols from the standpoint of a subjective human. We are constantly in dynamic cycles of codetermination with the milieu. On the one hand, the milieu shapes and nurtures human communities and relational individual, namely, acts as a matrix. On the other hand, the milieu is shaped through the traces that individuals and communities leave throughout their lives, namely, their imprints. The diverse milieus we are living in today have been built through different historical processes that span beyond the individual life.

The medial matrix surrounding an individual is shaped through the historical process of leaving collective medial imprints (the whole of the individual imprints), and, in turn, shapes individual human beings. I highlighted three main processes at play in this nurturing process of individuals by the medial matrix. Individuals make sense of the world and themselves in a participatory way, what Hanne De Jaegher refers to as participatory sense-making. They draw meanings and narratives from the cultural imaginary in which they are thinking and which limits what they can think like the transparent walls of a fishbowl. Their actions are constrained and guided by socially accepted practices. Interconnected and mutually supporting sets of practices compose social structures that organize individual behaviours and actions in each specific milieu.

Desirable medial imprints are ethical actions. Regardless of one's intentions, awareness, knowledge, etc., the individual leaves imprints in the milieu that can be (1) direct (all the significant traces left by the action of an individual), (2) domino effect (spanning up to the global level) and (3) mediated (by virtue of being a member of a social structure). I consider as significant traces any effect that is important enough to cause irreversible damages (such as a species extinction) or to trigger domino effects resulting in changes of the milieu or of the environment affecting other species'survival and other humans' health. Mediated medial imprints are the significant effects that an individual agent has by virtue of being a member of a social structure. There can be consequences of one's existence (market analysis), one's being somewhere (shame, imagination), one's appearance (body, clothes) and social roles (body language), one's inaction (letting something happen and not complying with a consensual practice), etc.

From the perspective of the phenomenological agent, most actions are done automatically without deliberation. The agent is blinded by automatisms, partial ignorance about the possible consequences, the transparent walls of the cultural imaginary, dogmatic rock-bottom beliefs hiding and deforming parts of the reality, and the social role endorsed at the moment of taking the action. On top of that, any deliberation process is bodily constrained and limited by time and by the theory of rationality the agent is relying upon. Nevertheless, an action can move from the automatic grey part of the automatic–deliberate spectrum to the centre of the spotlight of deliberate action, especially when the agent faces unexpected outcomes, social pressure and blame, and moral shocks.

All in all, the individual agent is always situated at a particular point in time and space. Her perceived surroundings are results of historical processes, and they shape and influence how she thinks, what she does and who she becomes. This is the milieu understood as a nurturing matrix shaping the individuals. The individual agent is also taking decisions and acting on her milieu, in other words, leaving imprints that, together with imprints of others who share the milieu, shape the milieu. This is the milieu taken as the result of collective human imprints. Finally, as the individual is both relational and the place of ethical agency, any action leading to a significant imprint on the milieu and on the environment can be considered a potentially harmful or ethical action.

Notes

1 Some aspects presented in this chapter were already mentioned in Droz (2018b, 2020).
2 The feminine is used as gender neutral.

Bibliography

Anderson, Benedict. 1983. *Imagined Communities: Reflections on the Origin and Spread of Nationalism*. London: Verso.
Balázsi, Ágnes, Maraja Riechers, Tibor Hartel, Julia Leventon, and Joern Fischer. 2019. 'The Impacts of Social-Ecological System Change on Human-Nature Connectedness: A Case Study from Transylvania, Romania'. *Land Use Policy* 89 (December): 104232. https://doi.org/10.1016/j.landusepol.2019.104232.
Bath, Frederik. 1969. *Ethnic Groups and Boundaries: The Social Organization of Culture Difference*. Universitetsforlaget, Bergen/Oslo: George Allen & Uwin.
Bayart, Jean-François. 1996. *L'illusion Identitaire*. Paris: Librairie Arthème Fayard.
Berque, Augustin. 1996. *Etre humains sur la terre: principes d'éthique de l'écoumène*. Paris: Gallimard.
———. 2000. *Ecoumène: Introduction à l'étude Des Milieux Humains*. Berlin.
———. 2000. *Médiance, de Milieux En Paysages*. Augustin Berque, Berlin, Paris, France. 2000
Besse, Jean-Marc. 2018. *La Nécessité Du Paysage*. Editions Parenthèses. Marseilles, France: Editions Parenthèses.
Bourdieu, Pierre. 1977. *Outline of a Theory of Practice*. Translated by Richard Nice. Cambridge Studies in Social and Cultural Anthropology. Cambridge: Cambridge University Press. https://doi.org/10.1017/CBO9780511812507.
———. 1990. *The Logic of Practice*. Stanford, CA: Stanford University Press.
Bragg, Elizabeth A. 1996. 'Towards Ecological Self: Deep Ecology Meets Constructionist Self-Theory'. *Journal of Environmental Psychology* 16 (2): 93–108. https://doi.org/10.1006/jevp.1996.0008.
Callicott, J. Baird. 1997. *Earth's Insights: A Multicultural Survey of Ecological Ethics from the Mediterranean Basin to the Australian Outback*. Los Angeles: University of California Press.

'Cambridge Dictionary | English Dictionary, Translations & Thesaurus'. n.d. Accessed 7 January 2021. https://dictionary.cambridge.org/.

Carter, Robert. 1996. 'Introduction to Watsuji Tetsuro Rinrigaku'. In *Watsuji Tetsuro's Rinrigaku: Ethics in Japan,* Watsuji Tetsuro, Translation: Yamamoto Seisaku and Robert E. Carter. Albany: State University of New York Press (SUNY).

Castoriadis, Cornelius. 1975. *L'institution imaginaire de la société.* Paris: Seuil.

Chakroun, Leila, and Laÿna Droz. 2020. 'Sustainability through Landscapes: Natural Parks, Satoyama, and Permaculture in Japan'. *Ecosystems and People* 16 (1): 369–83. https://doi.org/10.1080/26395916.2020.1837244.

Chater, Nick, and George Loewenstein. 2016. 'The Under-Appreciated Drive for Sense-Making'. *Journal of Economic Behavior & Organization,* Thriving through Balance, 126 (June): 137–54. https://doi.org/10.1016/j.jebo.2015.10.016.

Colding, Johan, and Stephan Barthel. 2019. 'Exploring the Social-Ecological Systems Discourse 20 Years Later'. *Ecology and Society* 24 (1). https://doi.org/10.5751/ES-10598-240102.

De Jaegher, Hanne, and Ezequiel Di Paolo. 2007. 'Participatory Sense-Making'. *Phenomenology and the Cognitive Sciences* 6 (4): 485–507. https://doi.org/10.1007/s11097-007-9076-9.

Deguchi, Yasuo, Jay L. Garfield, Graham Priest, and Robert H. Sharf. 2021. *What Can't Be Said: Paradox and Contradiction in East Asian Thought,* 1st edition. New York: United States of America: Oxford University Press.

Di Paolo, Ezequiel, and Hanne De Jaegher. 2017. 'Neither Individualistic nor Interactionist'. In *Embodiment, Enaction, and Culture: Investigating the Constitution of the Shared World,* 87–105. Cambridge: MIT Press.

Drengson, Alan, Yuichi Inoue, Arne Naess, and Gary Snyder. 1995. *The Deep Ecology Movement: An Introductory Anthology.* Berkeley, CA: North Atlantic Books.

Droz, Laÿna. 2018a. 'Cross-Cultural Environmental Ethics and Activism in Japan and Taiwan"'. *The Proceedings of the International Conference on Multicultural Democracy,* May 10–13th 2018, 281–92.

———. 2018b. 'Watsuji's Idea of the Self and the Problem of Spatial Distance in Environmental Ethics'. *European Journal of Japanese Philosophy: EJJP* 3: 145–68.

———. 2020. 'What Ethical Responsibilities Emerge from Our Relation with the Milieu?' In *Human and Nature,* edited by Arto Mutanen and Mervi Friman. Turku: Turku University of Applied Sciences, 15–30.

———. 2021 (forthcoming). 'Review of "Watsuji on Nature" by David W. Johnson'. *The Journal of Japanese Philosophy.*

Fournier, Marcel. 2012. 'Bourdieu, la raison et la rationalité'. *Cites* 51 (3): 115–28.

Gaard, Greta. 2011. 'Ecofeminism Revisited: Rejecting Essentialism and Re-Placing Species in a Material Feminist Environmentalism'. *Feminist Formations* 23 (2): 26–53. https://doi.org/10.1353/ff.2011.0017.

Gaard, Greta, and Lori Gruen. 'Ecofeminism: Toward Global Justice and Planetary Health'. In *Blackwell Philosophy Anthologies,* edited by Andrew Light and Holmes Rolston. Oxford: Blackwell Publishing, 276–93.

Geertz, Clifford 1993. *The Interpretation of Cultures: Selected Essays.* London: Fontana Press.

Gergen, Kenneth J. 2011. 'The Self as Social Construction'. *Psychological Studies* 56 (1): 108–16. https://doi.org/10.1007/s12646-011-0066-1.

Gibson, James. 1979. *The Ecological Approach to Visual Perception.* Boston, MA: Houghton Mifflin.

Gopnik, Alison. 2011. *The Philosophical Baby: What Children's Minds Tell Us about Truth, Love & the Meaning of Life.* London: Random House.

Grimshaw, Jean. 1986. *Philosophy and Feminist Thinking.* Minneapolis: University of Minnesota Press.

Hanspach, Jan, Jacqueline Loos, Ine Dorresteijn, David J. Abson, and Joern Fischer. 2016. 'Characterizing Social–Ecological Units to Inform Biodiversity Conservation in Cultural Landscapes'. *Diversity and Distributions* 22 (8): 853–64. https://doi.org/10.1111/ddi.12449.

Harris, Graham. 2007. *Seeking Sustainability in an Age of Complexity.* New York: Cambridge University Press.

Haslanger, Sally. 2015. 'Distinguished Lecture: Social Structure, Narrative and Explanation'. *Canadian Journal of Philosophy* 45 (1): 1–15. https://doi.org/10.1080/00455091.2015.1019176.

———. 2016. 'What Is a (Social) Structural Explanation?' *Philosophical Studies: An International Journal for Philosophy in the Analytic Tradition* 173 (1): 113–30.

———. 2018. 'What Is a Social Practice?' *Royal Institute of Philosophy Supplements* 82 (July): 231–47. https://doi.org/10.1017/S1358246118000085.

Heberlein, Thomas A. 2012. *Navigating Environmental Attitudes.* Oxford, New York: Oxford University Press.

Hermans, Hubert J. M. 2011. 'The Dialogical Self'. The Oxford Handbook of the Self. 10 February 2011. https://doi.org/10.1093/oxfordhb/9780199548019.003.0029.

Imanishi, Kinji. 2013. *A Japanese View of Nature: The World of Living Things by Kinji Imanishi.* London: Routledge.

Jasper, James M. 1997. *The Art of Moral Protest: Culture, Biography, and Creativity in Social Movements.* Chicago: University of Chicago Press.

Johnson, David W. 2019. *Watsuji on Nature: Japanese Philosophy in the Wake of Heidegger,* 1st edition. Evanston, IL: Northwestern University Press.

Labadi, Sophia. 2012. *UNESCO, Cultural Heritage, and Outstanding Universal Value: Value-Based Analyses of the World Heritage and Intangible Cultural Heritage Conventions.* Lanham, MD: AltaMira Press.

Lamarque, Peter. 2007. 'On the Distance between Literary Narratives and Real-Life Narratives'. *Royal Institute of Philosophy Supplements* 60 (May): 117–32. https://doi.org/10.1017/S1358246107000069.

Larsen, P. Bille, Philippe Le Billon, Mary Menton, José Aylwin, Jörg Balsiger, David Boyd, Michel Forst, et al. 2020. 'Understanding and Responding to the Environmental Human Rights Defenders Crisis: The Case for Conservation Action'. *Conservation Letters* https://doi.org/10.1111/conl.12777.

Lévi-Strauss, Claude. 2008. *Oeuvres.* Paris: Gallimard.

Liederbach, Hans P. 2001. *Martin Heidegger Im Denken Watsuji Tetsuros Ein Japanischer Beitrag Zur Philosophie Der Lebenswelt.* Iudicium, Taschenbuch.

Linstead, Alison, and Joanna Brewis. 2004. 'Editorial: Beyond Boundaries: Towards Fluidity in Theorizing and Practice'. *Gender, Work & Organization* 11 (4): 355–62. https://doi.org/10.1111/j.1468-0432.2004.00237.x.

Loizidou, Elena. 2013. *Disobedience Subjectively Speaking. Disobedience.* Philadelphia: Routledge. https://doi.org/10.4324/9780203796962-16.

MacCormack, Carol, and Marilyn Strathern, eds. 1980. *Nature, Culture and Gender,* Illustrated edition. Cambridge, New York: Cambridge University Press.

Manfredo, Michael J., Tara L. Teel, Michael C. Gavin, and David Fulton. 2014. 'Considerations in Representing Human Individuals in Social-Ecological Models'.

In *Understanding Society and Natural Resources: Forging New Strands of Integration across the Social Sciences*, edited by Michael J. Manfredo, Jerry J. Vaske, Andreas Rechkemmer, and Esther A. Duke, 137–58. Dordrecht: Springer Netherlands. https://doi.org/10.1007/978-94-017-8959-2_7.

Marietta, Don E., and Hedberg Maps. 1995. *For People and the Planet: Holism and Humanism in Environmental Ethics*. Temple University Press.

McCarthy, Erin. 2010. *Ethics Embodied: Rethinking Selfhood through Continental, Japanese, and Feminist Philosophies*. Plymouth: Lexington Books.

Ménage, Gilles. 1650. *Dictionnaire Étymologique de La Langue Française*. Chez Biasson: Paris. https://gallica.bnf.fr/ark:/12148/bpt6k507912/f1.item

Merchant, Carolyn. 2005. *Radical Ecology: The Search for a Livable World*. New York: Routledge.

Mestre, Mireia, and Juan Höfer. 2020, December. 'The Microbial Conveyor Belt: Connecting the Globe through Dispersion and Dormancy'. *Trends in Microbiology*. https://doi.org/10.1016/j.tim.2020.10.007.

Naess, Arne. 1973. 'The Shallow and the Deep, Long-range Ecology Movement. A Summary'. *Inquiry* 16 (1–4): 95–100. https://doi.org/10.1080/00201747308601682.

Neisser, Ulric. 1991. 'Two Perceptually Given Aspects of the Self and Their Development'. *Developmental Review*, The Development of Self: The First Three Years, 11 (3): 197–209. https://doi.org/10.1016/0273-2297(91)90009-D.

Northoff, Georg. 2014. *Unlocking the Brain: Volume 2: Consciousness*. Oxford, New York: Oxford University Press.

O'Doherty, Kieran C., Alice Virani, and Elizabeth S. Wilcox. 2016. 'The Human Microbiome and Public Health: Social and Ethical Considerations'. *American Journal of Public Health* 106 (3): 414–20. https://doi.org/10.2105/AJPH.2015.302989.

O'Neill, John, Alan Holland, and Andrew Light. 2008. *Environmental Values*. New York: Routledge.

Ostrom, Elinor. 2015. *Governing the Commons: The Evolution of Institutions for Collective Action*. Canto Classics. Cambridge: Cambridge University Press. https://doi.org/10.1017/CBO9781316423936.

Ota, Kazuhiro. 2018. 'Critical Cosmopolitanism and Fudo Theory: Tetsuro Watsuji and Alfred Schütz'. *Comparative Studies* 45: 73–81.

Oxford English Dictionary. 2000. Oxford: Oxford University Press.

Palomo, Ignacio, Carlos Montes, Berta Martín-López, José A. González, Marina García-Llorente, Paloma Alcorlo, and María R. G. Mora. 2014. 'Incorporating the Social–Ecological Approach in Protected Areas in the Anthropocene'. *BioScience* 64 (3): 181–91. https://doi.org/10.1093/biosci/bit033.

Plumwood, Val. 1991. 'Nature, Self, and Gender: Feminism, Environmental Philosophy, and the Critique of Rationalism'. *Hypatia* 6 (1): 3–27.

———. 2002. *Feminism and the Mastery of Nature*. New York: Routledge.

Rawls, John. 2005. *A Theory of Justice: Original Edition*. Cambridge, MA: Harvard University Press.

Rozzi, Ricardo. 2018. 'Biocultural Homogenization: A Wicked Problem in the Anthropocene'. In *From Biocultural Homogenization to Biocultueral Conservation, Ecology and Ethics 3*, edited by Ricardo Rozzi, Roy H. May Jr., F. Stuart Chapin III, Francisca Massardo, Michael C. Gavin. Irene J. Klaver. Anibal Pauchard, Martin A. Nunez, Daniel Simberloff. Cham: Springer Nature, 21–47.

Rozzi, Ricardo, F. Stuart Chapin III, J. Baird Callicott, Steward T. A. Pickett, Mary E. Power, Juan J. Armesto, and Roy H. May Jr. 2015. *Earth Stewardship: Linking Ecology and Ethics in Theory and Practice*. Cham: Springer.

Rupprecht, Christoph D. D., Joost Vervoort, Chris Berthelsen, Astrid Mangnus, Natalie Osborne, Kyle Thompson, Andrea Y. F. Urushima, et al. 2020. 'Multispecies Sustainability'. *Global Sustainability* 3. https://doi.org/10.1017/sus.2020.28.

Sala, J. F. Aguirre. 2015. 'Hermeneutics and Field Environmental Philosophy: Integrating Ecological Sciences and Ethics into Earth Stewardship'. In *Earth Stewardship: Linking Ecology and Ethics in Theory and Practice*, 235–47. Cham: Springer.

Seed, John, and Joanna Macy. 2007. *Thinking Like a Mountain: Towards a Council of All Beings*. F New Edition Used. S.l.: New Catalyst Books.

Seibt, Johanna. 2020, Summer. 'Process Philosophy'. In *The Stanford Encyclopedia of Philosophy*, edited by Edward N. Zalta. Metaphysics Research Lab, Stanford University. https://plato.stanford.edu/archives/sum2020/entries/process-philosophy/.

Sevilla, Anton L. 2017. *Watsuji Tetsurô's Global Ethics of Emptiness: A Contemporary Look at a Modern Japanese Philosopher*. Global Political Thinkers. Palgrave Macmillan. https://doi.org/10.1007/978-3-319-58353-2.

———. 2018. 'Cultural-Moral Difference in Global Education: Rethinking Theory and Praxis via Watsuji Tetsurô'. *Educational Studies in Japan* 12: 23–34. https://doi.org/10.7571/esjkyoiku.12.23.

Strawson, Galen. 2015. 'Against Narrativity'. In *Narrative, Philosophy and Life*, edited by Allen Speight, 11–31. Boston Studies in Philosophy, Religion and Public Life. Dordrecht: Springer Netherlands. https://doi.org/10.1007/978-94-017-9349-0_2.

Taylor, Peter J. 2015. 'The Ethics of Participatory Processes: Dynamic Flux, Open Questions'. In *Earth Stewardship: Linking Ecology and Ethics in Theory and Practice*, 325–37. Cham: Springer.

Veyne, Paul. 1971. *Comment on Écrit l'histoire Suivi de Foucault Révolutionne l'histoire*. Paris: Point Seuil « Histoire ».

Vidal de la Blache, Paul. 1903. *Tableau de La Géographie de La France*. Paris: Hachette.

Von Uexküll, Jakob. 1909. *Umwelt Und Innenwelt Der Tiere*. Berlin: J. Springer.

Wang, Zhen, Marko Jusup, Lei Shi, Joung-Hun Lee, Yoh Iwasa, and Stefano Boccaletti. 2018. 'Exploiting a Cognitive Bias Promotes Cooperation in Social Dilemma Experiments'. *Nature Communications* 9 (July). https://doi.org/10.1038/s41467-018-05259-5.

Warren, Karen J. 1990. 'The Promise and Power of Ecofeminism'. *Environmental Ethics* 12 (2): 125–46. https://doi.org/10.5840/enviroethics199012221.

Watsuji, Tetsurō. 2004. *Milieu (Fuudo, Ningengakuteki Koosatsu)*. Tokyo: Iwanami Bunko.

———. 2007. *Ethics (Rinrigaku), First Published between 1937 and 1949*. Tokyo: Iwanami Bunko.

———. 2011. *Fūdo, Le Milieu Humain, Commentary and Translation by Augustin Berque*. Paris: CNRS.

Winters, Andrew M. 2017. *Natural Processes: Understanding Metaphysics without Substance*. Palgrave Macmillan. https://doi.org/10.1007/978-3-319-67570-1.

Yuasa, Yasuo. 1987. *The Body: Toward an Eastern Mind-Body Theory*. Trans. Nagatamo Shigenori and Thomas P. Kasulis. Albany: SUNY Press.

3   Sustainability

To ethically orientate our actions and projects that have impacts on the environment, we need to define the general direction of what outcomes are desirable and what outcomes should be avoided.[1] This brings us to the philosophical question of what is good, or at least, what is preferable. Our goal here is not to provide a comprehensive ontological and cosmological theory of the good but to reach some minimalist basic premises that could reach consensual agreement among diverse worldviews and cultural backgrounds. Pragmatically, we need to have a general direction of what is good in order to coordinate and collaborate to take common actions towards it.

In the introduction, I proposed two minimalist premises that could be consensual enough among individuals from different cultures. The first premise is "We value human existence". I mentioned that this premise opens the question of what is actually valuable in human existence, and that answers to this question are generally deeply rooted in rock-bottom beliefs depending on the worldviews and cosmologies adopted by the individual. Yet, we can quite safely assume that valuation of human existence is not limited to currently living human beings but includes the possibility of continuation of human existence in the future.

Then, to ask the question of "what is good" at the three different levels of the individual, the milieu and the environment may shed more light on a consensual ethical direction for common actions. At the environmental level, valuable human existence translates in basic survival and the fulfilment of the vital needs of (at least some) living human beings, that is, the survival of the human species. The easiest way to insure the possibility of fulfilment of the vital needs of human beings, both in the present and in the future, is to preserve a relatively healthy environment whose self-regulating processes are still autonomous enough to continue providing vital services without requiring any human intervention. Next, the concept of milieu emphasizes the importance of relations. What is good at the level of the milieu appears to be healthy relations, but how to develop such healthy relations remains mysterious. It raises plethora of questions of fairness, equity, distribution of resources and powers, etc. Finally, what is good at the level of the individual could be the possibility of leading a meaningful life.

What seems to be good at one level might conflict with what appears to be good at another. For example, to develop fairer relationships, distribution mechanisms and usages of the land at the level of the milieu might conflict with what some privileged and dominant individuals understand as their flourishing life. "Fairer" here is primarily in terms of human relationships, but could include other-than-human living beings and other entities. Moreover, to preserve the environment to provide sufficient resources for the needs of future individuals might require to limit the growth of the human population, which is likely to conflict with the dominant values of many milieus.

At the same time, the three levels are inseparable and intrinsically imbricated. Indeed, as the individual is constantly in cycles of codetermination with her milieu and with other individuals sharing the milieu, it does not take a big step to understand that a web of healthy relations around the individual supports greatly the possibility of leading a meaningful life. Moreover, the importance of the environment is also encapsulated as conditions for the flourishing of the individual, as most milieus value strongly some elements of the environment and as the environment is a provider of vital needs for human beings. These reflections bring us back to the second premise briefly mentioned in the introduction: "A healthy environment and a meaningful milieu are necessary conditions for human existence".

A direction for positive ethical actions is then to maintain and develop the conditions of possibility of meaningful lives, the most basic conditions of possibility being a healthy environment and a meaningful nurturing milieu. I refer to this direction as "sustainability". In this chapter, I defend my working definition of sustainability as the maintenance and the transmission of a living meaningful milieu and of a healthy environment. Consequently, actions and projects are non-sustainable when they are undermining the conditions for their long-term continuation and fulfilment. As such, they can be considered harmful. Sustainability gives us a direction relatively to which we can judge the goodness and the harmfulness of actions and projects.

First, I present and defend my working definition of sustainability that focuses on preserving the conditions of continuation of flourishing and self-determining human existences, for which it appears necessary to maintain the global environment healthy and to develop diverse and adaptable milieus. Second, I draw some normative implications from my definition of sustainability, namely, two criteria of evaluation of practices (the possibility of continuation and the minimization of harm), a necessary and precautionary prohibition and some virtues of holistic conduct. Finally, I discuss some objections to and limitations of this definition of sustainability.

3.1 Precaution, diversity and environmental autonomy

Without even speculating about future generations' desires and needs, we can already look at the world right now and observe a high diversity of worldviews and values concerning what is good in human existence. On top of this diversity in worldviews come differences in the applications of these worldviews to the real world. As Jamieson writes: "conflicts can occur not only among plural values, but even when we seek to apply a single value in different circumstances" (Jamieson 2008, 168). Even if we consensually agree on a general idea of the "good" as "sustainability", diverse interest groups and interpretations will differ regarding what concrete applications it implies. This diversity is irreducible and even desirable to a certain extent, as it fosters dialogue and the construction of better solutions.

Finally, there are more answers to the question of "what matters in human life" than living human beings. Accepting the plurality of worldviews is not only a precautionary step, as I discussed in the introduction, but it is also a necessary one if we believe that what makes human life so beautiful and precious is precisely the fact that each of us chooses how to build our life. Of course, any choice is informed and shaped by the milieu and the multiple interactions with others. But still, the individual remains an agent who thinks, imagines, chooses and acts, in short, who leads her life.

A possible way out of this high diversity of opinion regarding what matters in human life is to protect the possibility for individuals to choose themselves to lead their life in what they judge to be a meaningful and fulfilling way within some general ethical limitations. Interestingly, the United Nations Millennium Ecosystem Assessment presents "*freedom of choice and action,* including the opportunity to achieve what an individual values doing and being" as a "precondition for achieving other components of well-being" (World Resources Institute 2005). Yet, despite being recognized both as a "precondition" and as depending on the other components of well-being (security, basic material for good life, health and good social relations), and despite being erected as such, this crucial element of human well-being is not problematized, defined or justified in the report. This is understandable, as a philosophical detailed justification for self-determination and freedom of choice is likely to face reluctance from some member states with diverse cultural backgrounds. Still, in the context of a philosophical argumentation for sustainability, I need to justify the premise of self-determination and to defend it against possible objections. Nevertheless, in our state of uncertainties and doubts emerging from the pluralism of worldviews, preserving self-determination could be a precautionary step. In other words, the argument goes:

1   What should be maintained is the possibility for self-determining flourishing human existences.
2   It requires a healthy environment for socio-ecological, epistemological, normative, historical reasons, in particular for political reasons as it neutrally provides fulfilment of vital needs.
3   It requires meaningful milieus, as they carry meanings beyond the individual life span and codetermine human identities.

Yet, (1) often conflicts with (2) and (3), because of the problem of distribution of resources and power, and the dynamics of oppression and domination. Plus, these dynamics of hierarchization and the diversity of systems of distribution are often central and meaningful characteristics of the social structure of a particular milieu.

Remarkably, diverse milieus (3) and human existences (1) are dependent on each other, and both depend on the global environment (2). Therefore, the order of the definition could be reversed and start with the maintenance of meaningful milieus, which entails and depends upon flourishing human existences, and thus requires a healthy environment. It could be understood or developed in a way that does not require human existences. For instance, in ecology, the idea of milieu could be adapted to any species. In this understanding, which differs from the updated concept of milieu used in this book, placing the premise of preserving meaningful milieus first could not necessarily entail the preservation of human existences. Nevertheless, because the idea of meaningful milieus appears to be more complex and debatable than human existences, the normative premise of preserving the possibility for flourishing human existences is kept here at the starting point, as it is likely to be simpler and more consensual.

These three propositions composing the starting point of my definition of sustainability echo the three levels of the individual, the milieu and the environment. At the level of the human individual, what is desirable is health and self-determination. These require a healthy environment. Finally, as a characteristic of human life is the need for meaning, human well-being also depends on meaningful and diverse (to foster the variety of choices) milieus.

3.1.1 Flourishing and self-determination

What are "self-determining flourishing human existences"? Self-determination is related to autonomy, which in turn opens an old normative and political debate. In the second chapter, I discussed how the individual is shaped by her milieu, as she is fundamentally relational. From such a perspective, the individual is never independent and fully autonomous from the webs of relations she is part of. Then autonomy is relational, as it refers to self-government, and the self is relational (Mackenzie and Stoljar 2000). Self-determination and autonomy, like freedom, are easier to define negatively, and Kymlicka gives us such a definition: "No particular task is set for us by society, and no particular cultural practice has authority that is beyond individual judgement and possible rejection" (Kymlicka 1991). In other words, the individual has the ability – in political theories of liberalism, this also means the right – to choose, change and act on all aspects of oneself. From a relational perspective, this goes beyond Kymlicka's definition to the idea that "oppressive social conditions of various kinds threaten those abilities by removing one's sense of self-confidence required for effective agency" (Christman 2018). Then, to be able to determine what one's wants to do with her own life, one needs, first and foremost, sufficient self-confidence to be able to reflect about it. But then, self-confidence rests most of the time on other's judgements and appreciations, and values and judgements about what is a good life are mostly borrowed from the cultural imaginary and the practices of other individuals in our community. Finally, as we have seen, we even make sense of the world together with others. To choose how to lead our life and engage in long-term projects (MacIntyre 2007), we also necessitate some degree of predictability of our surroundings, which depends on our state of knowledge, as we discussed previously. Then, what is left of self-determination seems quite fragile.

Nevertheless, any individual is still an agent, and she codetermines her milieu and its practices as much as she is shaped by them. That is, from her specific situated standpoint, she is still reflecting on and choosing to engage in some practices that she negotiates with her community. Even if practices and social structures are relatively stable, they are still the changing results of continuous adaptive processes including actions and decisions by numerous individuals (Macioce 2012). For an individual to be able to reflect on her situation and to make relatively informed choices about her own life, some basic conditions are necessary. Ideally, she needs to enjoy a relatively stable health, and to have the possibility to adapt herself to changes in her surroundings.

Health was broadly defined by the World Health Organization as "a state of complete physical, mental and social well-being and not merely the absence of disease or infirmity" (Roger Few 2013, 144). Out of the three types of well-being distinguished here, physical well-being is the most basic. The absence of pain and minimizing it when it cannot be avoided can be considered as the most basic element necessary for the individual to lead a flourishing life. The World Health Organization defines in turn mental well-being as a state in which "every individual realizes his or her own potential, can cope with the normal stresses of life, can work productively and fruitfully, and is able to make a contribution to her or his community". Out of these four points, the realization of one's potential reflects ambiguously the question of individual self-determination. Second, the capacity to cope with stress echoes the importance of adaptability. Third, the capacity to work is linked with the possibility of sustaining oneself. Fourth, the possibility to contribute to one's community fits with our account of the relational individual, and with the idea of social well-being appearing in the definition of health.

From the perspective of the observer, a minimal hedonism is a simple and consensual starting point to judge what is a flourishing human existence. It is hardly debatable that minimizing pain is, a priori, a good thing. In turn, we can define "harm" as the infliction of pain to a sentient being, starting with human beings. But leading a life merely deprived from pain could hardly be characterized as flourishing. The flourishing of human life necessitates meaning. And if meanings are created together with other humans, inspired by the cultural imaginary and shaped by the milieu, they are nonetheless subjective. In other words, the observer can hardly judge what is meaningful to the phenomenological agent. The proponent of moral minimalism Virvidakis Stelios writes:

> at the end of the day, one might embrace scepticism about the existence of objective criteria for the correct appraisal of better or worse individual styles of life, precisely to the extent that they are unique and could be compared to artistic performances which cannot be repeated.
>
> (Stelios 2014)

This scepticism urges us to take the highest precautions when judging what counts as a meaningful life from the perspective of the observer. The maintenance of the conditions for self-determination of what is a meaningful life for the phenomenological agent appears as a basic yet crucial precautionary step.

Then, what should be maintained and nurtured is the individual capacity to survive, thrive and flourish without crucially depending on the "good will" of other dominant human beings. Indeed, if the conditions of flourishing entirely depend on the choices and opinions of an observer agent, it places the latter in a position of strict domination that can lead to abuses, and undermine self-determination and flourishing of the dependent individuals. Conflicts between what is subjectively judged as good for oneself and what is evaluated as good by the observer emerge daily. Self-harm is an archetype example. On the one hand, basic needs or vital needs are usually consensually agreed upon, as shown by conventions such as Human Rights. On the other hand, individual sacrifices of those needs for a higher end are not uncommon. Stories of sacrifices for the sake of spiritual beliefs, coherence or purpose can be found in any milieu, from Christian martyrdom to Japanese kamikaze. Moreover, what is "good" for someone is ever-changing, so we need to preserve the possibilities for multiple adaptations to different unexpected circumstances.

In sum, to maintain the possibility for self-determining flourishing human existences, we need to prevent at least two types of harm affecting, first, human health, and second, the quality of life for human beings. Nothing surprising here. But what about harm affecting mainly the nonhuman natural environment? In order to justify protecting it for the sake of maintaining the possibility for flourishing human existences, we still need some justifications. Indeed, the destruction of elements of the nonhuman natural environment that does not directly affect the health or the quality of life of human beings could be value-free. Then, on what grounds do we value elements of the natural environment on which we do not depend directly?

3.1.2 *Maintaining the global environment healthy*

First and foremost, valuing requires valuers, so nothing can be said to be good "out of relation to human existence, or at least to some consciousness or feeling" (Sidgwick 1874). I do not argue here that nature has value *by itself*, intrinsically and regardless of our human standpoint. As the concept of milieu shows, we are always situated and embodied in our subjective human existence, and defending intrinsic values that do not depend on human beings valuation would dangerously deny our limited and largely ignorant standpoint. In Goodin's words:

> Good of nature can only be realized through interactions with human consciousness is not to say that nature is « good » merely because it is, in some crassly material way or even in some deeply spiritual way either, « good for » the human beings involved.
>
> (Goodin 1992, 45)

This being said, to argue for the attribution of intrinsic values to some elements of our milieu by human communities and groups is not ruled out, and can completely fit in particular worldviews. All what I deny is the objective existence of such values outside of a human perspective.

One might doubt how valuing the global environment worth maintaining contributes to the flourishing of individual human lives. Valuing objects "does not easily translate into harms and benefits to the valuer" (Jamieson 2008, 11). On the contrary, when it comes to meanings, it is not uncommon to value an element of our milieu that is, in fact, contributing to harming us. From a certain standpoint, this is the case with private cars and air pollution, or unhealthy yet tasty food. So, even if most of us agree that human existence is somehow valuable, some would disagree that preserving the natural environment is a condition for the flourishing of human existence. They could argue that the progress of technology will surely allow us to develop substitutes for most of what we originally received from the environment, and to build human-made technological systems to provide us with clean air and water – this view is called technological optimism. Some could even argue that with climate change, humans are already manipulating the whole natural system of the planet, and that it is just a matter of learning how to control it to maximize human benefits. By using substitutes and skilfully manipulating the provisioning systems, we could live at least as well as we did when we fulfilled most of our needs by taking resources directly from the natural environment without making much large-scale human changes to them. In other words, such a completely human-controlled world would not be anyhow less desirable than today's world. Some would even argue that such a world would be instead much safer than a world of wilderness.

This kind of argumentation often emerges as a result of an underlying dualism between nature and humanity that opposes them both. But even if we acknowledge that humans are part of nature, and that interactions between humans and their environment are natural by definition, then why should we worry about making important changes in our natural environment. Species have always disappeared, and now human beings are the dominant species and eradicate other "weaker" species, but why should that bother us, if that is precisely the normal evolutionary course of natural phenomena.

Holland points out the "paradox" of natural capital, which means that "the realization of its potential is at one and the same time the limitation of its potential" (Holland 1999, 56–64). He notes that natural capital can be said to contribute to welfare directly by virtue of its characteristics, by being beautiful, interesting or symbolic, as constituting a special sphere or place of operation, and for its function, such as a sink. But most of the time, natural capital contributes to welfare indirectly by being transformed to provide material for production and consumption. Then, the very usage of the capital by human beings is often destroying or at least affecting the capital itself, because none of it can be "used" without intervention. Eating food is an example, and tourism for environmental wonders follows the same scenario. Flocks of tourists rushing to see a specific natural phenomenon in a particular ecosystem can lead to the destruction of that ecosystem. But some might ask: Why should we care about it? The "consumption" of the natural capital by humans as the dominating species might be completely "natural".

Several replies can be made to this argument. First, the currently available knowledge establishes that we are vulnerable to changes in nature and that we directly depend on the environment. The more we learn about how our health is intertwined with the environmental surroundings, the more it appears clearly that we cannot dismiss the dependence of human health on nature. The health conditions of all organisms in an ecosystem are interconnected, up to the scale of the planet (van Bruggen et al. 2019). Microbial communities connect ecosystems globally, through successive habitats via food webs (Mestre and Höfer 2020). Microbes of animal origin sometimes "spill over" to human beings, due to contact among wildlife, livestock and human beings, become zoonotic diseases that affect human health, and can lead to global pandemics. Pandemic risk is increasing, driven by human unsustainable use of the environment, wildlife trade, agricultural expansion, etc.: "The underlying causes of pandemics are the same global environmental changes that drive biodiversity loss and climate change" (IPBES 2020, 2). Some – including several United Nations agencies (e.g. UNEP and ILRI 2020) – argue now for a "One Health" approach that brings together disciplines and recognize that environmental health, animal health and human health are closely interlinked (Rock and Degeling 2015; Verweij and Bovenkerk 2016). In other words, we need to care for the health of the global environment, because it is the foundation of our human health. As the case of pandemics shows, the impact of environmental destruction is not limited to individual health, but spans up to public health at the regional and global levels, which affects inevitably societies, cultures and milieus. Pragmatically, we need to restrict and limit human activities that lead to environmental harm to protect our health, our milieus and the projects we care for in this world.

A second reply draws from the symbolic and meaningful importance of the interdependence and inseparability of humans, milieus and the environment. Individuals are constantly in cycles of codetermination with their milieu, and a large part of what they care about in the milieu is closely entangled with the natural environment. It might be possible to imagine a totally artificial milieu, a milieu with no access to any potentially uncontrollable natural elements. Still, insofar as where we are standing now, such a milieu would still be supported by some natural processes, even if those are beyond the knowledge of the individual. Moreover, the wild natural environment is highly likely to still be a part of the cultural imaginary of the individual. To imagine a world or lifestyle without these two major links with the natural environment (physically supporting human lives and meaningfully inspiring human imaginations) would bring us too far from the present, and be irrelevant to our discussion, as our goal here is to draw ethics for ourselves rooted in our current existence.

Defending the protection of the natural environment because it is meaningful to us is common. For example, Robert Goodin argues that what we value is coherence and purpose, and these emerge from situating our life in a larger context such as the natural environment:

> If what we value about nature is that it allows us to see our own lives in some larger context, then we need not demand that that nature be *literally* untouched by human hands. We need demand merely that it have been touched only *lightly* – or if you prefer, *lovingly* – by them.
>
> (Goodin 1992, 53)

Dale Jamieson agrees that "context is very important to the character of our experience" (Jamieson 2008, 158). According to him, just as we value authenticity and rarity, "we value what is natural because we value nature's autonomy" (ibid, 166). Nevertheless, a limitation to this argument is that what makes the flourishing of a human life is highly subjective and strongly determined by the meanings of the local milieu. So coherence and purpose might not be necessarily valuable for an individual life in every milieu. Claiming that the value of the environment precisely comes from the fact that it provides us with a larger context for our experience seems to be situated already in a particular worldview.

A third reply draws from the limitedness of our capacities and time to act and think, and of our inescapable ignorance, given the fact that, even if all the knowledge about our supporting natural system would be available, no individual human being could master it enough to make informed decisions. These are all aspects reflecting the second characteristic of the milieu, which is that milieus are experienced by phenomenological agents trapped in our limited human and sociocultural standpoint. To have the capacity to control the whole system would then put us as individuals in an extremely precarious situation, as we would have the responsibility not to do any harm. The weight of this overwhelming tremendous responsibility is highly likely to paralyse our moral agency, and to push us into developing strategies to avoid taking responsibility. Then, to leave the environmental systems autonomous healthy appears to be the easiest and safest option, as we would not risk making terrible mistakes in their management or suffocate ourselves with responsibility. In other words, we need the environment to be autonomously healthy because, by doing so, we do not have to worry about it as it lies beyond the scope of our agency and responsibility.

A fourth reply has to do with the distribution of powers and knowledge in such a hypothetical world in which humans (as a group) would be in control of the parameters of their environment. If human beings were in complete control of their provisioning surroundings, that is, these could not sustain themselves without human intervention, then this state of affairs would give tremendous power to the individuals who are in charge and have the knowledge of managing such huge machinery, not only on the other species but also on the other human beings. Indeed, because milieus are interconnected globally in a dynamic network, as expressed by the third characteristic of the milieu, to give the power to disable others' milieus and ways of life to a few dominant human groups appears risky. There would be no escape for individual humans to go away and lead their own lives independently of a social system, which could be exploitative. The possibility for individual human beings and small groups to be autonomous and independent, allowing them to freely, that is, not constrained by other individuals, lead their life rests on the environment providing them with vital conditions that do not depend on human-made and controlled mechanisms. A quick look at the history of humanity tells us that oppressive and exploitative systems are recurrent and that crimes against humanity committed in the contemporary era do not convince us of the contrary.

Fifth, we can value the environment as irreplaceable, not because technology could not clone and artificially reproduce similar elements, but precisely because most natural environmental elements came into being without human intervention. Goodin writes: "at least some things produced by natural forces are irreplaceable, precisely because they have a history of having been produced by those natural forces. The things might be replicated artificially. But history cannot be so replicated" (Goodin 1992, 72). The currently dominant paradigm of ecology supports such a claim. Indeed, ecological systems are understood as non-linear, context-sensitive and feeding on variability "complex adaptive systems". According to Harris, they are very sensitive to unanticipated anthropogenic perturbations:

> Chaotic solutions do not converge to equilibrium; on the contrary, they diverge continuously so that very small changes (often indeterminably small changes) in the initial conditions lead to quite different outcomes. Such systems are irreversible: you cannot wind the clock backwards after a period of divergence.
>
> (Harris 2007, 54)

The point they make is that, be it natural processes or social processes, history is irreversible. By virtue of this historical irreversibility, any element of the environment can be regarded as precious, as they are irreplaceable.

Such a worldview where dynamic changes are the rule and not the exception turns upside down a worldview of stability and equilibrium. There is no "normal" or "original" equilibrium state of nature towards which aiming restoration efforts and orientating our actions and policies. This aspect echoes the fourth characteristic of the milieu, which is that milieus are continuously changing. Then, in such a worldview of dynamic changes, what should we maintain? We have seen a family of replies arguing that we must maintain the autonomy and health of (at least some) environmental processes. First, environmental health must be preserved, at all scales, because it is closely interlinked with and supports animal and human health. Second, the natural environment should be maintained because it carries meaning, and because diverse natural elements provide imagination with more diverse meanings. Third, the global environment should be maintained as an autonomous safety net for multiple ways of life, including other species, that is, a bigger context of ourselves and a limit to our agency and responsibility. Fourth, it should be maintained as a self-sustaining neutral provider of resources regardless of the other human beings judgements, such as human conflicts over land and water. Fifth, it should be maintained by virtue of its historical irreplaceability.

The second reply, namely, to keep the natural environment in its diversity as a carrier or meanings indicates a slightly different argument from the maintenance of environmental autonomy. What we should maintain is not only the environment as a chemico-physical receptacle, but also the diverse meanings built historically and intersubjectively and imprinted on the milieu.

3.1.3 Diverse and adaptable milieus

From a global perspective, milieus are highly diverse. This diversity is seen both as a value towards which we should design our actions – we should "protect" diversity – and as the *de facto* result of respecting other human beings' and communities' self-determination – we should "respect" others in their diversity. If we accept the idea that what should be maintained is the possibility for self-determining flourishing human existences, then diversity amounts to providing multiple varied options from which the individual (or the community) could pick up some and lead her life flourishingly. The protection of diversity is a tantamount to the development of better conditions for self-determination. As I argued, self-determination is desirable because we embraced precautionary pluralism and we rejected the existence of one single true worldview to be imposed on the world unequivocally. Then, preserving and cultivating diversity means maintaining a diversity of options from which we can choose and compose our worldviews. In other words, to have access to this diversity of options gives individuals the possibility to exercise their agency, which is an essential part of what it is to be human. Thus, diversity is not only the result of respecting other currently living human beings' self-determining agency, but it is also desirable to be maintained to allow future generations to exercise self-determination.

We can hear discourses drawing beautiful pictures of biodiversity, ecological diversity, cultural diversity, etc. For instance, the UNESCO Universal Declaration on Cultural Diversity links cultural diversity to human rights, freedom of expression, media pluralism, multilingualism, etc. and states that "cultural diversity is as necessary for humankind as biodiversity is for nature" (Article 1). Yet, when we look closer at it, the concept of "diversity" is poorly defined. The concept of diversity has been discussed for almost a century in ecology, where debates regarding how to measure ecological diversity are still raging (Sarkar 2010). Some argue that the number of different types of entities in a system, namely, richness, is a measure of diversity. But then, if there are 30 types of trees in my forest, but if the large majority of trees are pine, and if there are only a few individuals of other species, then diversity could be judged to be lower than if there is less disparity in the relative abundances of individuals. Others might say that rarity is a measure for diversity, which could be interpreted as the presence of rarer entities, or of entities that are more restricted in geographical range, or even of entities that have unique features. These questions take an even more complex dimension when we realize that depending on where we set the borders of the area to be assessed, the picture changes dramatically.

The political debates around conservation have urged scientists to come with biodiversity measures that are quantifiable and precise, to be able to compare and prioritized between different usages of the land (for habitation, agriculture, etc.). This need for clear communicable and comparable values conflicts with the complex reality of the field. Estimations of biodiversity, like estimations of pollution, are calculated based on sampling methods that are not anodyne insofar as their scales and sampling grids influence the results, as well as the real-life limitations regarding exactly where and when the samples are taken. Very different conclusions can be drafted from the same set of data, depending on the scale and the indicators that are chosen for the analysis. On top of this come political usages and choices regarding sciences. Instrumentalization of knowledge is inevitable, especially for environmental sciences, as environmental assessments are at the heart of conflicts of interests. "Who holds the power to simplify complexity" holds then a tremendous, yet often invisible and ignored power, as simplification is an essential step of communication, which is necessary to decision-making (Pascual et al. 2013, 164). Moreover, the evaluation of our ignorance and the opinions about how to deal with uncertainties differ greatly. Graham Harris describes this phenomenon as "epistemological relativism: people do have other values and beliefs – and views about the treatments and weights to be placed on the uncertainties – which clash with those of the science community" (Harris 2007, 238).

Whatever the precise meaning and evaluation criteria for biodiversity, there is a wide consensus on the importance of its protection. The ground for this consensus is not the intrinsic value of (humanly constructed) species or similar arguments as those made previously for the protection of the global environment. Instead, it is the pragmatic idea that "a species, or other element of biodiversity, has option value when its continued existence retains the possibility of future uses and benefits" (Faith 2019). The possibility that biodiversity elements might contribute to human welfare or even be necessary for human survival in the future seems enough to justify present protection of those potential assets. The relevance of this claim is as obscure and uncertain as the basis of what biodiversity is, but it remains a compelling argument, precisely because the limitedness of our current knowledge urges us to adopt a highly precautionary posture.

Assessments and discussions about the concept of diversity related exclusively to species and environmental elements are already highly complex and politically charged. But species and ecosystems literally cannot complain about the choices made by the scientific community to assess and analyse them. When it comes to cultural and linguistic diversity, one can imagine the situation becomes even more complex. Nevertheless, discourses advocating the protection of cultural diversity became ubiquitous and popularized by the United Nations Educational, Scientific, and Cultural Organisation for the last decades (Skutnabb-Kangas, Maffi, and Harmon 2003). Among these discourses, Sarkar distinguishes two lines of argument. First, cultural and linguistic diversity might geographically coincide or even be causally linked to biodiversity. The argument goes: "even if this geographical coincidence is no more than a result of historical contingency, the protection of cultural diversity would contribute to the protection of biodiversity", and vice versa (Sarkar 2010, 134). The ideas of biocultural habitats (Rozzi 2015) and of milieus capture this intertwinement between biological and cultural diversity. Indeed, one can imagine that different technics of usage of the land and dependence on different crops will inevitably contribute to supporting different types of ecosystems. In this argument, the protection of cultural diversity is justified by its instrumental effect on the protection of biodiversity.

Second, cultural diversity can be considered valuable, because of its "transformational power" or "adaptational strength" (Mühlhäusler 1995). This brings us back to the argument for maintaining the conditions for self-determining flourishing human lives spelled out in the beginning of this section. In Sarkar's words: "Maintaining cultural and linguistic diversity keeps a variety of future options open for the human species as a whole" (idem). Then, a decrease in cultural or linguistic diversity would decrease the "common pool of knowledge" to respond to environmental challenges. Moreover, interactions between different cultures might generate more reliable knowledge than otherwise. In other words, more intra- and inter-diverse milieus would increase human adaptability, and then improve the environmental survivability of human species. At the level of the milieu, adaptability translates into social resilience to brutal socio-political and environmental changes.

The idea that adaptability is a necessary condition for our survival is almost tautological. The above argument that diversity fosters better adaptability echoes the common sense saying not to "put all your eggs in the same one basket". The underlying intuition says that higher diversity leads to better adaptability and better chances of survival and flourishing. This assumption also underlines my previous argument about the need for the environment as a neutral non-partisan provider. We need a global autonomous and healthy environment to avoid the homogenization of cultural diversity into some dominant worldviews and oppressive social structures, because this diversity of lifestyles is a safety net, and a consequence of respecting the autonomy of other individuals, groups and beings. A global homogenization of ways of life leads to a diminution of biological and cultural diversity of habitats, which, in turn, reinforces the establishment of some life habits and worldviews. This is the problem of biocultural homogenization, which captures "feedback processes entailing interwoven losses of biological and cultural diversity" and mutually reinforcing worldviews and practices, ways of life and habitats, and the habits of different species sharing a habitat, including human communities (Rozzi 2018). To avoid the vicious cycle of biocultural homogenization, we need to maintain biocultural diversity at different scales within a global autonomous and healthy environment.

But this line of argument soon encounters an important obstacle, which is that the protection of diversity of lifestyles might conflict with some other intuitions regarding self-determination, namely, the idea to give "equal" opportunity to individuals to freely design their own life. Putting aside egalitarianism and questions of distributive justice, we still face the problem that in most milieus, social roles are closely linked with hierarchical positions of domination and power. As we discussed in Chapter 2, these social roles and positions in the social structure and the milieu an individual agent belongs to are crucial components of her identity. Identities are relationally built, and domination is a very common type of relation between human beings. Many of these relations come with the circumstances the agent is born into or living in. They constrain and limit the self-determination importantly, and can even produce harm. But at the same time, they give meaning to one's position, role and even existence in the milieu. Indeed, many meanings carried by the milieu are intertwined with justifications for domination and the distribution of powers and resources. These patterns of distribution create differentiated positions and social groups that share highly different lifestyles and develop different subcultures. In short, some level of inequality produces differences and preserve diversity of meaning, while simultaneously restricting the range of individual self-determination.

Then it is not sufficient that milieus carry diverse and adaptable meanings and practices, but it is also essential to broaden the access to these and to tools to negotiate individual positions within the webs of relations in order to foster self-determining flourishing individual lives. Another helping point is that milieus are not fixed in time. They are changing and adapt to circumstances. This adaptation is crucial for sustainability. Some meanings and practices will have to be abandoned, such as the symbolic association of dark air pollution with economic development, while other new meanings and practices will need to be developed. This dynamic process of critical evaluation and change of meanings and practices of the milieu is continuously ongoing. We can then suppose that equipping individuals with critical tools to understand their situation and constraints, evaluate them and choose how they want to lead their life from the range of options available to them is a key. Yet, utmost precaution must be taken about this educational claim, as ways of reasoning and making sense of the world are diverse and situated. The attempt to educate everyone according to what one dominant community judges to be critical thinking might lead again to the imposition of this dominant worldview and the homogenization of others' milieus.

All in all, sustainability is a precautionary multi-level dynamic process, which continuously redefines its objects. That is why the answer to the question "what should we maintain" is composed of "possibilities". Peter Jacques gives the following definition:

> Sustainability is the imperfect process of building and maintaining global social systems of capable, accountable, adaptive, just, and free people who can make important decisions and trade-offs with foresight and prudence and who foster the robust, self-organizing, dynamic, and complex ecosystems around the world for now and future generations.
>
> (Jacques 2020, 19)

Without buying his assumptions about "global" social systems of "just and free" people, his definition still gathers the most important elements we have discussed, namely, the fact that sustainability is the process of maintaining self-organizing and robust ecosystems according to the precautionary principle.

Not only the environmental processes and the relational individuals are continuously changing and dynamic, but also the milieus. The dynamicity of the concept of milieu does not only fit the latest views of the worlds as imbricated non-linear systems, but it also justifies sustainable changes. Indeed, the milieu is not a fixed image grounded on local lifestyles and values from the past. Instead, it is an ever-changing web of meanings that are adapted by the currently living individuals to better fit their needs and values. Then, nowadays, if it is necessary and urgent to renounce to some seemingly old traditions and replace them with more sustainable practices, it should not be considered as a regrettable loss or a threat to a supposedly stable harmony, but as expressions of the ongoing cyclic relationships between living individuals and their long-lasting milieus. This echoes claims about "cultures of sustainability", which, "as soon as they crystallize into fixed states, closing their boundaries and fixing their borders, they risk losing their elasticity and porosity", precisely that gives them the power to adapt and survive changes (Kagan 2010, 1100).

3.2 Normative implications

Based on our working definition of sustainability, we can draw some normative implications. Indeed, if our working definition of sustainability might

Table 3.1 Working definition of sustainability in terms of milieu

---

Working definition of sustainability

Sustainability is the maintenance of the conditions of possibility of continuation of (1) self-determining flourishing human existences. It entails (2) maintaining the general processes of the global environment healthy to limit the possible harmful consequences of the conflicts of distribution and domination, and (3) cultivating meaningful, diverse and adaptable nurturing milieus.

---

give us a general direction towards which orient our ethical actions, it appears to be still insufficient to assess what outcomes are desirable and what outcomes should be avoided, and more importantly to judge harmful actions and projects, and to assign moral responsibility. This is the object of this section. Before entering into details, let us restate our working definition of sustainability (Table 3.1).

From this definition, I will first argue for two criteria of evaluation of actions, practices and projects, namely, the minimization of harm and the possibility of continuation. Second, I will develop a necessary and precautionary prohibition that stems from this definition of sustainability, that is, the prohibition from destroying the global environmental systems to the point where these systems cannot independently provide healthy living conditions to current and future human beings. Finally, I will draw some practical implications at the individual level, in the form of virtues of holistic conduct.

3.2.1 Criteria of evaluation: possibility of continuation and minimizing harm

Practically, we need a criterion to evaluate what actions, practices and projects we shall engage in and which we shall avoid and change. I argue for a double criterion of evaluation. Actions, practices and projects should be evaluated according to (1) the types and levels of harms produced – in other words, their noxiousness – and (2) the possibility of continuation of the practice in the long term. Unsurprisingly, these two criteria are closely linked. As we will see, non-sustainable actions and projects are undermining the conditions for their long-term continuation and fulfilment, and as such, they can be considered harmful. Let us explore first the harmfulness of projects and then move to what entails the criteria of possibility of continuation.

It would be a mistake to take for granted the commonsensical principle "first, do no harm". It is not widely respected, as individual agents keep taking actions that harm others, and most worryingly, many of these actions are deliberate. Let us clarify first what "harm" means. Harm has a living sensitive object, who, as a result of harm, suffers. It can be a disturbance of a state of health, and it can be linked to the infliction of pain. A widely used definition of pain is "an unpleasant sensory and emotional experience associated with actual or potential tissue damage or described in terms of such damage" (Bonica 1979). Pain is usually transitory but can indicate long-term damage.

Without entering clinical definitions of nociceptive, neuropathic and psychogenic pains, it is clear that pain can come from different sources. Harm refers to the relation between the source of the pain or damage who can be an agent, and the sufferer. Not only human, but animals and other living sensible organisms can experience pain, leaving aside the question of drawing a clear border between what counts as a sensible organism. Moreover, as environmental, animal and human healths are interconnected, inflicting harm might have unexpected repercussions far beyond the entities that experience the direct pain or damages.

There can be different types of harm. Jamieson distinguishes three relatively to the sufferer, namely, harm threatening human health, affecting the quality of life of human beings or affecting mainly nonhuman elements of the environment (Jamieson 2008, 10). The first type of harm affecting human health can be traced back to pain and death, but is restricted to an individual human being. The second one can be much more wide-encompassing, and relates to the idea of flourishing discussed before. As I showed before, if we understand human flourishing as relying on meaningful milieus and on the possibility for self-determination, then harm affecting nonhuman elements of the environment could be understood as one type of harm affecting human quality of life. Yet, as the latter is still highly subjective, it is relevant to keep an independent category for harm affecting nonhuman environmental elements. This is especially important as improving the human quality of life might come along with inflicting serious harm, pain and damages to nonhuman living beings. Eating animal meat is an example of such a conflict, as the human pleasure and nutrient satisfaction come to the price of animal suffering and death.

Following this reasoning, I add a fourth type of indirect harm affecting elements of the milieu. This type of harm is indirect insofar as the elements of the milieu are not *all* living sensible beings, and at least, they are not considered as such when seen as elements of the milieu. Damaging elements of the milieus inflicts pain and suffering indirectly, by touching human individuals and groups who care and attach some importance to these elements of the milieu. For instance, the destruction of some monuments recognized as of cultural importance, such as World Heritage monuments, is harmful insofar as it inflicts strong emotional pain on many human beings, and can deeply damage the webs of meanings with which they make sense of the world and affect their worldviews. Despite the high levels of distress that might be felt and affect human individuals' lives and flourishing in the long run, the destructed monuments are not experiencing any pain or suffering.

The four types of harm can be found together in many environmental problems. For example, high levels of water pollution in a river can threaten directly the health of the members of the local community who rely exclusively on this water to drink and wash themselves. Besides, river water pollution is likely to affect the quality of life of other human groups, more distant or downstream, who used the river for playing, fishing and celebrating some traditional rituals. River water pollution also affects nonhuman environmental elements such as fresh water organisms and countless local species relying on a clean fresh water stream. Finally, high levels of water pollution also amount to damages to elements of the milieu. Indeed, the river might have been believed to host some local gods and spirits, or might have been celebrated for its remarkable beauty, or might be a central place appearing in legends and tales shared across generations. Some specific species of fishes or birds relying on the river might also be essential elements of the identity of some people and might play a central role in the worldviews of some others. In more abstract and general terms, it could be said that the harm affecting an apparently independent nonhuman environmental element – the river – cascades to the milieu, affecting in turn the quality of life of human beings, and finally threatening their flourishing and health. If such a clear succession is not the rule, it shows that these four types of harm can be overlapping and closely interconnected.

The sources of these different types of harm are multiple and complex, but they often involve more or less directly human actions, as I discussed in the last chapters about medial imprints. Some actions, projects or lifestyles of individual human beings produce multi-scaled harmful consequences like in the example of the river. Yet, it might be difficult, or even impossible in our current state of knowledge, to trace clearly the causal relation between the agents inflicting the harm and the sufferers. We will discuss this problem in Chapter 4: Responsibility. However, this difficulty is not a reason to renounce the infliction of harm as a criterion of evaluation for current actions, lifestyles and projects. Indeed, as phenomenological agents, we are all continuously confronted to the uncertainty regarding the possible outcomes of our actions and lifestyles. But this does not prevent us from evaluating our actions, lifestyles and projects regarding their potential multi-scaled and multi-type harm to the best of our current knowledge.

It is easier to evaluate and predict harms affecting human health and human quality of life. In many regions, the prediction and evaluation of these types of harm is also regulated by laws and is often at the heart of most compulsory and legally binding "environmental assessments". Harms affecting the nonhuman environment are harder to estimate for the average agent and require expert scientific knowledge, which is why they are more and more integrated in these "environmental assessments" that are required in most national and regional laws before starting any large-scale project. For the individual phenomenological agent, such expert evaluations are inaccessible, forcing us to rely on second-hand information and advice from the scientific community, the states and non-governmental organization.

Finally, when it comes to harm affecting the milieu, the evaluation of the potentially harmful consequences of one's action is even more complex, and is usually not taken into consideration. That is because the milieu is simultaneously subjectively perceived and shared, so what appears to be significant and important to me might not be from the perspective of my neighbour.

Similarly, my neighbour might be devastated by the destruction of some elements of the milieu that barely caught my attention. Nevertheless, the loss of some elements of the milieu, for example some landscape items, is likely to deeply affect almost all the inhabitants of the area (and possibly beyond), especially if those elements are important landmarks serving as a point of references structuring the shared worldview transmitted through education and public discourses. Some of those most consensual and historical elements are recorded on heritage lists and enjoy relative legal protection.

As we live and develop ourselves in codetermining cycles with our milieu, it comes as no surprise that harm towards some significant elements of our milieu can deeply harm ourselves. We invest tremendous time and energy in projects that we estimate meaningful, and that, in turn, give meaning to our life and actions. Some individuals dedicate their life to particular artistic projects or scientific enterprises, others work restlessly for the success of their business. Others still might aim seemingly more humbly at the survival and transmission of some values, habits and lifestyles that lie at the core of their worldview of a meaningful human existence. All in all, we usually highly value the projects we dedicate our life to develop, including many elements of our lifestyles. And because of the importance we give to those, we suffer from damages affecting them, even if our health or our quality of life might not be directly affected by those apparently external damages to things and ideas "outside" of us.

A crucial difference between this type of indirect harm related to damages to the milieu and the other three types of harm is that the former is not restricted to living sensitive beings. Again, of course, a milieu necessitates living human beings to exist, as it is composed of meanings and values that cannot exist without a valuer, that is, a phenomenological agent. But a milieu carries meanings beyond the individual life span. For most of us, the survival and success of meaningful projects (understood as including lifestyles, religious traditions, political ideals and artistic craftsmanship) beyond our individual death matters. It is not uncommon for individuals to dedicate so much energy and time to the success and development of some particular meaningful projects that it comes at the expense of one's health and wealth.

On the one hand, as phenomenological agents, we are stuck in our specific standpoint situated in space and time, and we have a limited possible life span. On the other hand, we draw meanings to make sense of our existence and actions from our milieu, that is rooted in the distant historical past, and that will survive our short individual existence. As we were shaped by a milieu transmitted to us by the ancestors who designed it by their imprints, future not-yet-born human beings will also be shaped by the milieu we are contributing to design right now. Future not-yet-born individuals are likely to get inspired by the projects and ideals to which we dedicated our present lives. They are also highly likely to orientate and make sense of their life by borrowing most of the meanings and patterns of lifestyles from what we are designing and preserving right now. It follows that the individual characteristics of who exactly will live in the distant future, a classical problem of inter-generational ethics, matters less than how they will live and if they will take over from us our most meaningful projects. From this perspective, future generations have full power over the endurance of presently valued projects and ideals (Droz 2019b).

This brings us to the second criteria of evaluation of current projects, actions and lifestyles, namely, the possibility for their continuation. Given the importance and values that are attributed to some projects and lifestyles, individuals are likely to suffer from their failure and collapse. However, some long-term projects running today undermine the conditions for their continuation and fulfilment. For instance, an economic project entirely reliant on fossil fuels is doomed to fall flat in a short-term future. Its meanings and symbols will collapse alongside the project and new meanings will need to be created and adopted. Such failure is not irrelevant, as it leaves social and cultural scars and can involve collateral damages such as social unrest and biodiversity loss. Thus, the sustainability of projects and worldviews we are inventing now is a forward-looking requirement.

Moreover, not all worldviews we are constructing today are equally sustainable, in the sense that some undermine the very conditions for their continuation. We can suppose that worldviews rooted on abusive mechanisms such as racism and sexism are deemed to stir protests and social instability. Hence, this second criterion focusing on the possibility of transmission of a meaningful historical milieu rules out several types of unsustainable projects, even if they might not appear to be immediately harmful. It constrains the actions of today's people in relation to future people. Finally, the high and irreducible complexity of our world's social and natural systems also urges us to follow a precautionary principle and preserve the highest diversity of sustainable elements of the milieu (at all levels, from biological diversity to the diversity of worldviews). Indeed, future people might understand similar elements of the milieu better or in other ways than us, and some of these elements may even become important conditions for their survival and flourishing as human beings (e.g. the potential of genetic diversity for the development of new medicine).

This second criterion of evaluation, namely, the possibility of continuation, is not centred on harm affecting living beings, but on maintaining a meaningful nurturing medial matrix. Transmission and maintenance focus on the conditions of possibility of continuation of the changing meanings and projects of the milieus. Together, these two criteria of minimizing harm and avoiding projects that cannot be continued in the long term allow us to sort out desirable projects and lifestyles. Like in the case of the evaluation of noxiousness, evaluation of the possibility of continuation of projects and lifestyles in the long run is highly complex and we cannot prevent some level of uncertainty. Nevertheless, we can still consider and judge actions, projects and lifestyles according to the best of our current knowledge, and following a strong precautionary principle.

3.2.2 Necessary and precautionary prohibitions

The double criteria of evaluation of projects and practices regarding their harmfulness and the possibility of their continuation in the long term, associated with the precautionary principle, lead us towards some general prohibitions. In line with our working definition of sustainability, it urges us to preserve a global healthy environment, for the reasons described in 3.1.2. In other words, human beings have, individually and collectively, a necessary prohibition from depleting or destroying the global environmental systems to the point where these systems cannot independently provide healthy living conditions to current and future human beings. Independently here means without significant human intervention on these systems.

Such a general prohibition inevitably raises multiple questions regarding its implications at different scales. To the best of scientific knowledge now, we are not able to fully model the flows and multi-scaled interactions at the global level. One cause of this difficulty might be irreducible, being the highly complex and hard to predict entanglement of complex and adaptable systems. Practically, coordination of human activities affecting the global environment is largely depending on nation-states and multinational corporations (who still must be subject to laws of at least some nation-states). Human agency is thus scattered, and there is no single individual or group of individuals who possibly could take some particular actions and answer to this necessary prohibition. The efficiency of the coordination of human activities under the territorial division in nation-states is relatively slow, especially compared to the ubiquity of transnational environmental flows that structure the global environmental systems, such as the water flows.

If we go down the scale of the global level of the environmental systems that we must protect, we soon encounter the multiple overlapping scales of the milieus and their diverse social structures. Currently, most usages of the land are organized and coordinated following different regimes of land property. Land property is fiercely guarded, at least by the land owners – or legal users, regardless of their specific designation – who sometimes fence their land and reserve the right to use their land as they please. They tend to be reluctant to comply with regulation or external advice about how to use – or refrain from using – the land they own. Implicitly, they seem to assert a "right to destroy". Interestingly, the Roman Law distinguished between three regimes of property, which are *usus* (the right to use a thing without deriving profit from it), *fructus* (the right to use a thing and to derive profit from it) and *abusus* (the right to alienate a thing by sales or material destruction). The latter is the etymological origin of "abuse", and is no longer existent in most contemporary legal regimes. Indeed, Goodin lists the main components that underline property rights in the contemporary Anglo-Saxon legal landscape, namely, a right to use – sometimes exclusively – and to transfer, and concludes: "there is no necessary, analytic connection between a right to destroy and any of those other more standard rights of property" (Goodin

1992). Moreover, there seem to be no theoretical justification to assert a right to destroy from a right to property, especially once we understand that the flows supporting the global and regional environmental systems cross private property borders, as well as they cross national borders.

It follows that the right to use (a piece of land, a natural resource, etc.) – which includes property rights – is subject to limitations following the two criteria of evaluation listed before. That is, property owners are banned from engaging in long-term projects that undermine the conditions for their continuation or fulfilment by destroying the self-regulating environmental systems of their land and other areas beyond the borders of their property. This prohibition is not restricted to property owners, but to any person in a position to use and significantly deplete the local land and its functioning and role beyond the local plot of land. As land owners are legally identifiable and liable for human interventions on their land, they are at the front line of such a general prohibition, even if they happen to be not individual human beings, but corporations, companies or nation-state.

Debates are likely to rage around what counts as depletion and destruction (e.g. Ostrom 2015). More precisely, the most precautionary postures regarding the preservation of the environment are likely to conflict with economic policies and economic interests of some individuals and companies. The main actors who are in a position to benefit financially from a project are often also likely to downplay the severity of the depletion. Given that financially dominant agents are often closely tied to the most powerful actors using the land without precautions for their own economic interest, they are in a position allowing them to influence public opinion. Their well-founded marketing strategies manipulate the public and political consensus-making regarding what can be expected from property owners and entrepreneurs towards minimal liability and economic policies based on exponential discounting. In this situation, the "public consensus" regarding non-destructive usage of the land tends to be orientated towards short-term profit, and neglects precautionary prohibitions.

The evaluation criteria "do no harm" reveals to be helpful to finding a way out from this pragmatic dead end. As we have seen, harm has living existing sensitive subjects, primarily other living human beings. Minimizing harm to human health and quality of life urges us to work on providing minimal welfare to all living human beings. Minimal welfare could be measured and provided following the idea of basic needs or the idea of capabilities, but the immense debate surrounding these questions goes beyond the scope of our argumentation here. Still, we placed the possibility of self-determining flourishing human existences at the root of sustainability, and we recognized that in the midst of the diversity of social organizational and sense-making patterns structuring milieus, inequalities of wealth and relations of domination often play a central role. Therefore, we cannot advocate intra-generational equalitarianism without overstepping on other milieus' webs of meanings.

Inequalities (such as the fact that a few individuals and groups of individuals own some land and have the rights to use it "as they wish", while most do not own any) do not inevitably conflict with our two criteria of evaluation of projects, as long as most sensitive living organisms, primarily other human beings, are not directly harmed from them. Yet, in the state of the real world nowadays, some human beings are *de facto* harmed by the consequences of some projects, and often they are harmed through the medium of environmental depletion crossing borders. So, if our two criteria do not rule out wealth and power inequalities, they still restrict the usage of these privileges.

The question is then: if projects and practices are judged according to the two criteria (no harm and the possibility of continuation) to the best of our current knowledge and applying the precautionary principle, would interrupting and banning the practices that significantly violate these two criteria be enough to ensure the preservation of healthy global environmental systems? Answers would rest on theoretical assumptions and technical measurement details of the realm of modelling experts, while an ultimate and definite reply could be provided only by applying them in the real world. For the sake of our argumentation, let us suppose that the extensive application of such criteria would be sufficient to ward off the danger of irreversible destruction of the global environmental systems.

Several subsidiary prohibitions can be drawn out from the general and abstract prohibition. A first subsidiary precautionary prohibition touches practices and projects leading to hardly reversible ecosystem degradation. Indeed, land degradation does not solely have a locally limited destructive impact, but, at the large scale it is occurring today, it also affects globally relevant environmental flows. Land degradation leads to biodiversity loss, threatens food and water security, can induce forced migration, and contributes to and is also exacerbated by climate change (IPBES 2018). Land restoration post-degradation is highly costly and complex, and it is uncertain whether the impacts on the global environmental flows can be restored. This subsidiary prohibition from depleting the local environment up to a temporarily irreversible state such as desertification or marine dead zones mainly regards land-use changes and agricultural and forestry practices for terrestrial ecosystems, and direct overexploitation for marine ecosystems, including fisheries and mining (IPBES 2019).

A second subsidiary prohibition is the limitation – or ban – of the mass production and spread of any element that might pose a threat to healthy living environmental conditions at the global scale. That includes the production of what is labelled as "hazardous wastes", which we do not know, to the best of current science, how to properly dispose and ensure their non-noxiousness in the long term. Nuclear wastes are also concerned by this precautionary prohibition, as they represent a global threat to the environmental systems and to healthy living conditions for current and future human beings. More generally, pollution – air, soil, water or else – is to be regulated and limited, as it directly impacts human health, while being a direct driver of biodiversity loss. Restricting human activities that contribute to climate change constitutes a third subsidiary prohibition. The interlinkages between climate change, land degradation, biodiversity loss, greenhouse gases flux and food and water security are well-known (IPCC 2019). In our interconnected world, global environmental processes are highly complex, tend to be mutually reinforcing and are influenced by direct and indirect drivers at various scales. Drawing subsidiary prohibitions from the general prohibition rests on scientific expertise. The three examples of subsidiary prohibitions given here are far from exhaustive, and are subject to changes and adaptations following the state of knowledge and the best available evidence.

These general and abstract precautionary prohibitions limit and constrain individual choices and behaviours. Before discussing how these general prohibition and the two criteria of evaluation translate into individually applicable guidelines for conduct, it is crucial to discuss a sensitive last point. The first prohibition did not concern only the provision by environmental systems of healthy living conditions to current human beings, but also to future human beings. Theorizing about future generations is thorny because it rests on debatable predictions. Brian Barry, who argues for sustainability as maintaining "equal opportunity across generations", suggests that most theories and discourses about sustainability rest on the assumption that "the size of the future population is no greater than the size of the present population" (Barry 1999, 93). Indeed, it is arguable that if the purpose of sustainability is to maintain the conditions of possibility of self-determining flourishing human existences, the number of individual human beings that can flourish on a single limited planet must be limited. Intergenerational equalitarianism advocates the transmission of "equal or better" conditions for flourishing of future human beings, assuming that the population growth will slow down and stabilize. Yet, without defending strict equalitarianism, we can argue for the transmission of the conditions of a minimum "decent society" (Beckerman 1999), and include in these conditions the necessity to relatively limit population growth correspondingly to the limited material environment we are relying upon. In my perspective, intra- and inter-generational egalitarianisms are not necessary requirements for sustainability, as it would likely overstep the respect for the diversity of organizational social structures of different milieus, both in the present and in the future. Though, it does not rule out the pursuit of equality of opportunity intra- and inter-generations in some milieus, as long as the projects and practices that would hypothetically come out from such an egalitarian enterprise do not conflict with the two criteria of evaluation and the general prohibition.

All in all, I argued for a necessary general abstract prohibition from engaging into any type of activity that threatens the preservation of the global healthy environment. This general prohibition is valid everywhere and anytime, regardless of borders and locally accepted worldviews. Indeed, the practices of a particular milieu organized around a worldview in opposition to this general prohibition would threaten other milieus. From this starting

point, I drew some subsidiary prohibitions proscribing the human-made production and spread of any element that might pose a threat to healthy living environmental conditions at the global scale (such as hazardous wastes
and nuclear wastes), practices and projects leading to hardly reversible land
degradation (such as some intensive agricultural practices) and unlimited population growth. Other precautionary prohibitions particular to the specific
environmental conditions and sociocultural practices of particular milieus
can be drawn from the premises discussed here, but I have neither the space
nor the expertise to discuss what particular precautionary prohibitions are
entailed by this argumentation for each particular milieu.

3.2.3 Holistic conduct

The necessary and precautionary general prohibitions and the normative criteria of evaluation have implications for individual agents. They normatively
restrict the range of options for actions from which an individual chooses in
order to lead a flourishing life. They also limit what can be considered in
some particular worldviews as individual freedoms and elements of individual flourishing. Nevertheless, this fact is not problematic, as it is precisely the
purpose of ethics to set limits to human endeavours.

However what could be considered problematic is the jump from a general
prohibition about the global environment to the conduct of an individual
agent. The two criteria that allow us to evaluate individual actions are important steps between the two scales. Yet, a further justification is needed.
Indeed, as we discussed in Chapter 2, many environmental problems are
caused by the cumulated effects of multiple actions, such as climate change.
This question is no stranger to ethics, as similar reasoning can apply with ly-
ing. If only I lie regularly, then it is likely that the society as a whole does not
significantly suffer from it. What is necessary is to apply the reasoning "what
if everybody acted as I do".

A famous application of this reasoning in environmental ethics is captured
by the concept of ecological footprint. The idea of ecological footprint relates
the human usage of resources to the carrying capacity of the planet. It develops a system of ecological accounting contrasting the biologically productive
area people use for their consumption with the area available, either in a
specific region or in the planet as a whole. Aiming at raising environmental
awareness, calculators that offer individual assessments of "how many planet
Earth" would be needed if everybody had the same lifestyle are widely available on the Internet. They rest on assumptions about stable world population,
and on the validity of the reasoning "what would happen if everybody acted
like me".

Multiple recommendations targeting individual behaviour emerge from
such a reasoning based on egalitarian assumption regarding the repartition of
equal parts of the earth capacity per individual living (which might be surprising, given that such egalitarian assumption is not common regarding the

distribution of financial wealth and other forms of capital). To fight climate change alone, individuals are then advised to cut down food wastes, switch to a plant-based diet, use public transportation, etc. The efficiency of these potential changes of behaviours is not a philosophical question, rather a scientific one regarding what human activities are the most harmful to nature. For this reason, I will leave the discussion of the details of these individual recommendations here.

Instead, let us develop the reasoning placing the possibility for self-determining flourishing human existences at the heart of sustainability. Two reasons were presented for preserving diversity: respect for self-determination in a multicultural diverse world, and the assumption that higher diversity leads to better adaptability, and thus increases not only the chances of survival, but also the opportunities for flourishing. From there, we can hypothesize that creativity and better stimulation of ideas increase the chances of finding the appropriate solution to a problem. One could further suppose that ideas and creativity are stimulated by living different experiences (physical, intellectual, emotional…). Then, individually, in order to reach a better adaptability, we would need to foster the diversification of experiences.

This argument for the diversity of experiences rests on several assumptions. First, there is no evidence that the development of creativity in *all* individual members of a population increases the chances of survival and thrive of this population. It might suffice that a few individuals develop their creativity in ways allowing them to find better solution to problems faced by their society. The underlying assumption here is the same as the one grounding the idea that self-determination is an important basis for flourishing, and thus desirable for each individual human beings, and not only for a few privileged. This egalitarian perspective on a sort of "human right" to self-determination is open to challenges and criticisms.

Second, one can doubt the assumption that the diversity of individual experiences fosters creativity and the development of original solutions to circumstantial problems. On the contrary, an individual experiencing very different ways of thinking and unexpected series of events might very well end up more confused than prompt to develop solutions. Still, given that we accept the importance of individual self-determination within the medial matrix, each individual must have access to a certain range of options and certain conceptual tools to develop her reasoning and choose what she does, that is, what she becomes, from the circumstances she is in. From the phenomenological agent perspective, adaptability is primarily social, as technologies, knowledge and sense-making are shared and participatory. To put it plainly, what might increase the individual chances of survival to brutal changes in her environment is nothing but collaboration with other individuals, even if it is only for the exchange or learning of technologies. Then, a certain social adaptability, that is, a flexibility and an ability to collaborate with others is a must. A socially adaptable individual constantly updates her knowledge and reassesses her actions and reasons for actions in reaction to the new pieces of information she gets from others and the world, including through media, books and the Internet.

From there, if we buy these assumptions, the argument for the diversity of experiences brings us a twofold normative conclusion related to individual behaviour, especially regarding the individual's attitude towards new pieces of information. On the one hand, a socially adaptable relational individual should be curious, namely, seek encounter and exchange with diverse unknown others. I will refer to this as the "virtue of curiosity". On the other hand, the individual must assess the new pieces of information and be ready to change her lifestyle, her habits and (part of) her worldview. Such an attitude entails avoiding dogmatic walls that prevent conscious change and exclude some experiences, including thought and conceptual experiences. I call this second normative conclusion the "virtue of adaptivity". Both sides of this twofold conclusion, curiosity and the absence of fixed walls, indicate a certain attitude and conduct of the agent towards other human beings, and towards new pieces of information in the world.

This curious and adaptable attitude towards conflicting or new pieces of information can seem to target more the epistemological attitude of the knower than the ethical conduct of the agent. Needless to say, both the epistemological attitude and the ethical conduct are closely linked – I would even say that the epistemological attitude is one component of the ethical conduct, as it shapes our understanding of situations in ways that influence our actions. In this sense, ethical conduct can be characterized as holistic as what might appear to be different parts of it are intimately interconnected and explicable only by reference to the whole of the holistic conduct.

In addition to the normative conclusions targeting the adaptable epistemological attitude of the agent, some other considerations closer to actions can be drawn from the double criteria of evaluation of projects. The criterion of minimizing harm appears often on the first line of the codes of conduct of different fields, such as the Hippocratic Oath and the idea "*Primum non nocere*" in medicine. It is also at the heart of many ethical systems, from the epicurean ethical hedonism to utilitarianism. Nevertheless, it is easy to slip from caring for others (starting with the other human beings, up to ecosystems) to harming others while meaning well, by way of imposing one's own needs on this vulnerable others. Indeed, caring can be an expression of power over another being. When we listed reasons to maintain the global environment healthy, the importance of humbly limiting our caring appeared as a clear necessity confronted with irreducible uncertainties regarding what would be beneficial for a complex "other" such as the global environmental systems. Because of this, it seems critical to cultivate a curious, opened and adaptable epistemological attitude regarding others' standpoints, worldviews and needs.

The criterion of the possibility of continuation of the projects and ideas also brings some important input to the holistic conduct of the individual. One could develop this criterion by saying that it is the temporal transposition of the spatially relevant idea "what if everybody does as I do". When it comes to personal conduct this continuation criteria translates into the importance of carefully choosing in what projects and worldviews we invest our emotions, beliefs, and trust. Curiosity is key, as it is impossible to evaluate and decide over the value of something unknown.

Simultaneously, what is known can affect directly one's worldview and that threat might lead some individual to refuse to learn some new pieces of information. As with caring, sharing information with others is not neutral, as it might influence them irreversibly. From the perspective of the well-intentioned giver (of information, or else), the question then becomes: Is there a way to give something without diminishing the agency of the receiver? Conversely, from the standpoint of the receiver, the careful and repetitive self-assessment and choice about the beliefs and projects structuring our worldviews and lifestyles might encounter some dogmatic walls in some rock-bottom beliefs.

If we understand self-determination normatively, as towards what an individual should thrive for throughout her life, then we have grounds to exclude some worldviews and arguments grounded on non-questioned rock-bottom beliefs. But as soon as we threaten some dogmatic worldviews and opinions with exclusion, we risk becoming dogmatic ourselves, and counterproductively exclude some individuals or groups of individuals whom we need to collaborate with for the sake of our global survival. This paradox of tolerance emerges frequently at attempts to pragmatically apply such abstract ideas in real-life conflicts.

Before attempting to draft a pragmatic solution to this apparent paradox, it is important to raise an objection to the whole argument of diversification of experiences. I call it the objection of homogenization, and it is grounded on the intuition that more interactions between different entities will lead to homogenization, and thus to less diversity. This fear is easily understandable from the previous description of interaction as being a process of mutual influence, leading to changes in all the parties involved, or at least in the most vulnerable parties in the case of an abusive relation of one-way domination. Fears of cultural homogenization – the reduction of cultural diversity – are also often expressed in reference to globalization, perceived as a form of unification force under the influence of some dominating actors. Whether this threat is realistic or not is widely debated, and to draw a clear answer to this political question is especially complex as soon as we take cultural changes as the normal course of events.

In contrast, in ecology, an increase in the number of interactions (both intra- and interspecies) is usually taken as fostering a better resilience of the ecosystem. Yet, a phenomenon of homogenization might still occur following some threshold effects. A reasoning inspired by these observations from natural sciences might be helpful here, by introducing the notion of threshold effect. Possibly, interactions between diverse individuals and groups of individuals increase social resilience, up to a certain threshold and relatively to the nature of the interactions. Regrettably, these are only speculations, and I do not have the expertise to further sort them out.

If we buy this objection of homogenization, or any other objections to the two problematic premises, then the argument for the diversification of experiences may come tumbling down like a house of cards. But if we accept this general line of reasoning, we end up with several principles to orientate individual holistic conduct. First, we mentioned the importance of cultivating curiosity by seeking enriching encounters and exchanges with diverse others. Second, we underlined individual adaptability as the readiness to change our lifestyle, habits and (part of) our worldview, starting with the avoidance of dogmatic walls that prevent conscious change and exclude some experiences. Third, we emphasized humility as the precautionary posture of refraining from imposing our ideas of needs and good on others. Finally, we came back to the crucial role played by self-reflection in the idea of individual self-determination. In relation to the continuation criteria of evaluation, we discussed how the individual can be expected to carefully choose in what projects and worldviews we invest our emotions, beliefs and trust.

3.3 Limits and priorities

This book aims to build a consensual framework on minimalist premises that can be widely accepted, because most of current environmental problems require common action for which we need to get as many people on board as possible. Still, all worldviews and projects cannot be accepted by an ethical framework, in which case it would make it null and void. Any ethics prohibits some behaviours and ideas that are deemed to be wrong in this particular ethic.

My working definition of sustainability contains three main points. It argues for the maintenance of the conditions of continuation of (1) self-determining flourishing human existences, (2) the general processes of the global environment healthy, and (3) the meaningful, diverse and adaptable nurturing milieus. I will discuss, in turn, what these three aspects of sustainability entail as limits and what they exclude, in conversation with possible objections emerging from proponents of some excluded opinions.

3.3.1 Objection to self-determination

In my working definition of sustainability, I made a swift move from the premise "we value human existence" to the maintenance of the possibility for self-determining and flourishing human lives. Some individuals and some communities (or community spokespersons and leaders) might object this move. Indeed, some might argue that complete faith and obedience from the group members in some texts or guidance of some leaders are the highest end of their conception of human life. They might plainly reject individual self-determination as a normative value. Simultaneously, they might as well accept self-determination at the group level, namely, that groups or communities should have the rights and freedom to determine by themselves – without precising the in-group decision-making processes that might range from inclusive and participatory to authoritarian.

This position rejecting individual self-determination is not uncommon. Many communities at various scales claim (through their community leader or spokesperson) to value traditions and full rule by an elite more than individual self-determination, as the latter can be perceived as leaving individuals too free, and thus easy target for maleficent ideas and influences. For example, heresy and apostasy among the majority Muslim community are considered serious threats in Malaysia, as the former (questioning the orthodox interpretation of Muslim doctrine) can ultimately lead to the latter, namely, the abandon of one's religious faith. Heresy and apostasy are presented as virus or dangerous illness that spreads "rampantly" and "can affect anyone, regardless of their social status and level of thinking" (Alwi et al. 2015). They are considered serious enough that "Faith Rehabilitation Centers" are available, to give an opportunity for the unlucky individuals to repent, and thus escape the death penalty that "must be carried out against apostasies due to heresy" according to "Islamic law".

As in the example of the Malay Muslim community, many individuals all over the world seem to hold the belief that one is born into a community withholding particular traditional worldview, practices and rules and thus must stick to it, refraining from questioning some of its aspects. Members of such communities are denied any self-determination *regarding some aspects of their life*, which can be more or less intrusive depending on the community. From this perspective, the idea of self-determination is perceived as a dangerous and wrong temptation. I will refer to these perspectives as fundamentalist worldviews.

Such fundamentalist worldviews brutally clash with my argumentation. Indeed, the questioning of one's own beliefs is a necessary step for self-determination, and it is echoed by the aspect of self-reflection in the holistic virtues I proposed. Similarly, to be free to change one's beliefs to others that one might estimate more fitting is considered essential in my framework as shown by the virtue of being adaptable. Thus, communities and milieus that are organized around the rock-bottom belief that the lifestyles and worldviews of their members must be imposed to them, monitored and controlled closely reject self-determination at the scale of the individual, and in turn, might reject my argumentation altogether. Community leaders might justify their will to impose and constrain other individuals in their assigned roles, and their expectation of full non-reflective obedience from the community members, by arguing that only these particular lifestyles and worldviews make human existence worth living for individual members of their community.

They might even go further and argue that my own argumentation is no different, insofar as it requires individuals to be curious, adaptable, humble and self-reflective, and excludes other ways of living. I could reply that the main difference is that my argument is to let individual explore many different possibilities of lifestyles and worldviews before choosing (and constantly reaffirming or being ready to change) the one that seems the most fulfilling and ethical to them. Thus, my argument does not exclude the possibility that after a self-reflective exploration, the individual decides to fit and obey to the role that was advised to her by her community, and there is no problem with such a choice insofar as it does not conflict with the general environmental prohibition and the two criteria of evaluation of lifestyles and projects. Fundamentalist proponents might not be satisfied with this reply, as it still denies them the relational right to impose on others some values and behaviours without any discussion.

It is here necessary to take a step back and remind ourselves the main purpose of developing this "consensual" framework, namely, the necessity to agree on taking common actions to tackle environmental problems. Self-determination was used as a key point in the argumentation for preserving autonomous environmental processes, but other reasons for preserving the global environment were also presented, such as the value of an healthy environment as setting a bigger context to our lives, and as being irreplaceable by virtue of its history. Then, it is possible for fundamentalist groups to refuse the premise of individual self-determination, while still considering crucial the maintenance of healthy global environmental processes for other reasons, because they provide a bigger context, or because of their irreplaceability, or still for other reasons particular to their own milieu and worldview that might have specific traditional and religious justifications for the preservation of nature.

Then, as long as the collective imprint of the fundamentalist community on the environment is not significantly harmful, exclusively focusing on environmental ethics we would have no reason to challenge their refusal of individual self-determination. Indeed, as long as the community is not violating the general environmental prohibition, external observers to the community would have no reason to intervene in the name of environmental protection to challenge the internal organizational structure and rock-bottom beliefs of the fundamentalist community. Nevertheless, if the community has a highly harmful collective environmental imprint (even if the harmfulness of this imprint mainly comes from the environmentally unfriendly lifestyles of a few of the community leaders), then external observers have an environmental ethical reason to criticize and maybe intervene towards fundamentalist community, as they significantly threaten the global environmental processes. "Significantly" here could be measured with egalitarian criteria of *per capita* imprints.

Such a position will likely strike more liberal minds as unjust, as it does not exclude the possibility for a fundamentalist community to have a few leaders leaving in highly environmentally costly lifestyles while keeping the large majority of their community members in extreme poverty. Indeed, an evaluation of what counts as "significant" environmental harm of the collective imprint *per capita* could even encourage the dominant community leaders to increase the extremely poor population of their community, as they would then receive more "per capita credits" to continue their own individual harmful lifestyles. The maintenance of global environmental processes might even be used by fundamentalist leaders to justify keeping their members in complete control and, in some case, in poverty.

A proponent of individual self-determination might argue that the collective project of such a fundamentalist community runs counter to the two criteria of evaluation. An organization structure that refuses self-determination and deliberation might seem harmful to some of its members. From the perspective of the external observer, torture and death penalty to individual members who challenge internal organization and authority might seem clearly unjust and harmful to these individuals. Moreover, a social structure lacking any inclusive participatory mechanism for minority claims (minority understood as political and social opposition) might be socially unstable, pushing some individuals to resort to violent militant methods to make their voice heard. In such a violent and unstable context, some citizens could get harmed in different ways, including harm to their physical integrity, in particular protesters and opponents to the regime. Resistance, sometimes violent, by opposing groups within the social structure could also hinder the continuation of such an authoritarian project. Thus, such a fundamentalist authoritarian social structure would be incompatible with both the no harm criteria and the criteria for continuation.

Despite the highest sympathy that I have for this argument, it is not hard to argue against it. A fundamentalist could argue that if control and education to obey are strong enough (imposed from childhood, or constantly monitored), then there would be no issue of continuation of the system. Moreover, it would be hard to establish what counts as harm, as individuals might not perceive harm in the same way as an observer would do. Individuals can internalize circumstances that are – or can appear – harmful to them, into becoming a central part of their identity. They can build their identity as victims and lack any desire to change this state of affairs because of fatalism or because they want to preserve a higher good, etc. External intervention, even limited to speaking to the victim and reminding her of her powers and probable rights, might cause significant damage and harm to the "victim" herself, as it would challenge her worldview and her identity.

A common way around this issue is to play with the scales, and argue that what matters is less *individual* self-determination, but more the *collective* self-determination at the level of the milieu – or the community. Then, what matters is that communities have the right to determine without external intervention the rules and norms by which they want to organize, including the decision-making systems by which these norms are decided. For example, if a community is organized around a strict reading of the Islamic Law, where designed religious interpreters of the sacred text rule in matters of ethics, justice and politics, no external observer has the legitimate right to challenge it.

The problem of heresy and apostasy (and their punishment by death penalty) is that if the observer has no right to criticize and intervene, and the members of the group neither, then the dominant individuals in the community are omnipotent over members of their community. One could imagine regulating mechanisms such as a committee of interpreters who must reach an agreement before any enforcement, making the organization slightly less despotic. But if we accept such a system, then it is hard to see how we could justify any intervention, as long as the global environmental prohibition *per capita* is not violated. The latter threshold being problematic as discussed above, especially because population number must be somehow regulated to answer to the global prohibition.

Will Kymlicka discussed "toleration and its limits" in the case of national minorities and suggested that some factors are potentially relevant to determine "the exact point at which intervention in the internal affairs", namely

> the severity of rights violations within the minority community, the degree of consensus within the community on the legitimacy of restricting individual rights [*which seem to require the possibility of discussing heresy*], the ability of dissenting group members to leave the community if they so desire [*apostasy*], and the existence of historical agreements with the national minority.
>
> (Kymlicka 1996, 169–70)

Importantly, Kymlicka discusses exclusively the case of national minority here, that is, minority who are already part of a national system of laws and rights, under which they can be judged by other observers within the nation. Internationally, or in the case of undocumented communities in areas beyond the rule of law of any state, the question of such "rights and laws" becomes relevant to international conventions and agreements (such as human rights), and the question of who might intervene is even thornier.

Nevertheless, "the ability of dissenting group members to leave the community if they so desire" seems to be a quite consensual and minimal starting point. Plus, as we will see in the chapter about responsibility, to have the possibility to dissociate oneself from a group and to have a minimal space for self-determination is crucial to preserve individual moral responsibility. In other words, to preserve moral responsibility (which is a key aspect in my framework), we need a basic requirement regarding the freedom to leave one's group. But accepting it as a norm in our "consensual" framework would exclude some fundamentalist proponents.

What seems to be required here is a prioritization of cases in which intervention can be legitimate regarding to our framework. The main goal of our framework is more basic than social justice, as it is to preserve the autonomous and healthy environmental systems. Then among the enormous whole set of practices, projects and policies that we need to review and assess, the intervention in other milieus to prevent and stop what we, as observers, perceive as harm against other human beings (which might not be perceived as harm from their subjective phenomenological perspective) falls relatively down in the priority list. Pragmatically, it is more important to get on board most communities, even the ones rejecting individual self-determination, so that all communities meet their "collective responsibilities" regarding the precautionary prohibition. Double standards might be necessary, with a stricter application of the criteria in one's own milieu than in distant other milieus. As I will discuss in the chapter about responsibility, beyond the general prohibition that is applicable globally, the priority for individuals' actions and interventions lies in their own milieus and social structures.

3.3.2 *Priorities, excluded projects and policies*

The highest priority for an ethic of sustainability is to comply with the necessary and precautionary prohibition from "depleting or destroying the global environmental systems to the point where these systems cannot independently provide healthy living conditions to current and future human beings". This general prohibition limits some specific human activities, such as practices leading to long-term ecosystems degradation, unsustainable overexploitation of ecosystems, the mass production of hazardous wastes such as nuclear waste, etc. It requires a global scale of judgement that cannot be reached solely by day-to-day life experiences of the phenomenological agent. It requires tools, investigation and collaboration to understand better such a global phenomenon and their regional and local variations. This is the key role of scientific research. Understanding what practices have severe consequences on environmental flows and systems necessitates scientific expertise, which rests on the collaboration of hundreds of scientists from different disciplines. Setting aside criticisms and problems within the scientific community (a critical problem being the financing of research projects influenced by industrial lobbies, and the censorship of some results), in general, it still thrives to provide data and to increase our understanding of the world.

   This also means that listing projects and practices that have the most harmful consequences and that may contribute to violating the general normative prohibition, *to the best of our scientific current knowledge*, is beyond my capacity, and would get us lost in the maze of details characterizing such an enterprise. What we are in a position to rule out here are some normative paradigms sometimes called "scientific", according to which many policies and projects are decided from the data provided by the scientific community. Many current dominant paradigms that claim to inform policy-makers "scientifically" while obscuring their own normative premises come from the field of economics.

The ethical framework I developed here supports a wide family of criticisms targeting the mainstream capitalist economic paradigm as the frame of reference to evaluate policies and projects. My two criteria of evaluation (minimizing harm and preserving the possibility for continuation) clash with the mainstream economic criteria focusing on economic growth. For the sake of quicker and higher monetary profits, the economic paradigm sets aside the questions of harm (often characterized as "externalities") and continuation. One of the main "scientific tools" used to evaluate projects and policies according to the economic paradigm is the idea of ecological exponential discounting (Müller 2013). Ecological exponential discounting is a discount function based on the idea that the value of environmental assets decreases exponentially with time. As a result, any policy or projects based on this idea are likely violating the continuation criteria.

Economic growth is often erected as the highest good, and a collapse of which would be the most serious threat to our well-being. Yet, in my working definition of sustainability, the highest value is the possibility of continuation of flourishing self-determining human existences, which is unlikely to be a function of economic growth. On the contrary, a focus on economic growth based on exponential ecological discounting undermines the conditions for sustainability.

Other related mainstream "values" brandished in the international political discourses are "industrialization" and "development", seen positively as the direction for success and improvement of the well-being of the population. Post-development theories and development criticisms are highly diverse (Escobar 1995; Latouche and Leung 1996; Rahnema and Bawtree 1997). Without developing a critique of these ideas, a quick look at the diversity of values and worldviews in different milieus is enough to put things into perspective. From the standpoint of my working definition of sustainability, there is no requirement for industrialization or "economic development". Indeed, local pre-industrial community lifestyles could be more sustainable in our sense. Health and well-being, often used as reasons justifying industrialization, do not necessarily increase with economic development. Indeed, economic development is regrettably often accompanied with severe environmental degradation (such as air pollution) that directly hinders health and well-being.

What is healthy for human beings is also very ambiguous and depends on what is valued in human existence, on the physical and mental vulnerabilities of the individual human in question and on the worldviews endorsed by her and her community. Hence, as industrialization and economic development are not, in their narrow sense, supporting any of the three points of sustainability (self-determination, global environmental processes, diverse and meaningful milieus), they cannot be erected as general direction for action. This is not to say that they are excluded, or that sustainability excludes economic development *de facto*. But it is to say that economic and developmental projects are subject to be evaluated regarding sustainability, as well as any other project. They are not exempt from being banned if they contravene to the general prohibition.

The highest priority is thus the respect of the general prohibition, which necessitates a global scale of judgement and wide collaboration to increase, develop and communicate scientific knowledge. But after excluding projects that contravene to the general prohibition, we are still left with a high variety of projects and lifestyles. The second priority is connected with one's own milieu and requires a local scale of judgement, along with the contextual application of the two criteria of evaluation of projects and lifestyles.

This second priority of shaping meaningful and nurturing milieus is connected with community resilience. In ecology, the meaning of the word "resilience" changed with the evolution of the discipline. It refers to the capacity of a system to survive and heal after a significant disturbance. Notably, it does not mean returning to a "pre-disturbance" state, as this is "most often *not* the most likely or even appropriate recovery strategy for human systems" (Wilson 2018, 91). Instead, it refers to the capacity of the community to adapt and preserve or regain some levels of wellness after a significant stress. For example, Norris et al. describe resilience as emerging "from a set of adaptive capacities – *community* resilience from a set of networked adaptive capacities" (2008, 135). They note that resilience "rests on both the resources and the dynamic attributes of those resources", that is, their combination. They advocate the diversity and distribution of the resources, as well as the network structure and linkages within the community, including community bonds, communication and collective decision-making and action. Interestingly, they suggest that their description of community resilience (that insists on inclusive and participatory decision-making processes) has nothing to do with "culture", because "we cannot envision a human culture or society in which the basic concepts of stress and disaster, resources, crisis, adaptation, and wellness do not apply" (idem, 145). Caution is required here, as other authors raise the critique that notions of resilience "are *reinforcing neoliberal pathways of development*" (Wilson 2018, 96), in other words, that they rest fundamentally on normative and culturally dependent premises.

This ongoing debate around community resilience illustrates the difficulties in which I bump into when attempting to clarify what "shaping meaningful and nurturing milieus" concretely means. It is the place for me to point out, again, that this aspect of sustainability requires a local scale of judgement. Doing so does not avoid the difficulties, but it importantly restricts their scope. Similar reasoning to the one presented as objections to individual self-determination can be applied here to argue against inclusive and participatory decision-making within the communities. Worries regarding cultural imperialism are not to be dismissed simply because "we cannot imagine" otherwise, as we are anyway trapped into the bowl of our own cultural imaginaries. Nevertheless, if we accept the value of individual self-determination, then it seems incoherent to reject *any* participatory mechanism in community decision-making, while it seems equally incoherent to normatively require individuals to actively make use of these mechanisms (at least at this stage of the argumentation).

What we can argue relatively safely – that is, without excluding from the start other worldviews – is that decision-makers who are in charge of evaluating the projects and actions of a given community, whoever they might be, should apply the criteria of evaluation of sustainability. Importantly, the application of these criteria does not exclude the application of other criteria specific to the milieu and the community. On the contrary, to evaluate and judge projects in one's own milieu regarding to the values and worldviews particular to this given milieu is essential for the preservation of diversity of meaningful milieus.

Figure 3.1 presents the flowchart of the priorities in the evaluation of projects according to the normative implications of my working definition of sustainability. I placed the question of the criteria of no harm before the question of the criteria of continuation, as the former is more common and less ambiguous than the second. More importantly, screening first according to the no harm criteria already confirms the acceptability of short-term projects that otherwise would be rejected because they are not possible to continue in the long term. Indeed, we have no reason to exclude short-term projects that are not harmful to any entities, and that do not contravene with the necessary and precautionary prohibition. For example, many art projects and festivals are ephemeral, and they take meaning and beauty precisely because they are ephemeral. Such projects contribute greatly to the dynamic web of meanings of the milieu, and it would be absurd to exclude them. Thus, the emphasis on the criteria of continuation is useful only insofar as it highlights potential future harm and other indirect and mediated harm that usually does not jump to the mind of decision-makers. In other words, the main advantage of the criteria of continuation is to highlight the importance of evaluation of the projects and lifestyles in a long-term temporal scale, but it does not rule out short-term harmless yet meaningful projects.

If the project is not ruled out at the end of this flowchart (if it does not end in a box "Stop it"), we can already be sure that the project does not contravene with the general prohibition, and that it is not highly harmful to other human beings. For example, a fossil fuel extraction project would already be ruled out as it harms current living human beings by contributing to worsen climate change and the increase of violent climate events leading to the loss of human lives. In case of doubt, if the project is judged not to be harmful to current human beings but might be harmful to other elements or to future human beings, then the criteria of continuation is crucial. If continuing the project in the long term significantly enhances the risks of harm, then it is equally ruled out. For instance, if a seemingly harmless practice could lead to a species extinction when continued in the long run, then it would be proscribed thanks to these two criteria taken together.

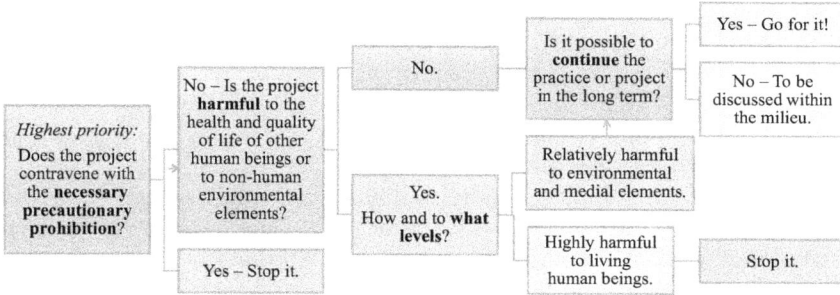

Figure 3.1 Flowchart of the priorities in the evaluation of projects

If the project is not harmful and it seems to be possible to continue the behaviour composing the project indefinitely, then it does not present any eliminatory element from the perspective of the ethics of sustainability I develop here. But if it might have limited harmful impacts, especially on the milieu, even if these harmful impacts result from the impossibility to continue such a project in the long run, then it becomes necessary to take the decision together with other members of the community – human, multispecies or else – sharing the milieu, as they will be affected by changes and losses in their milieu.

This brings us to the third level of priority, that is, immediate relations and how do we interact and develop consensus and agreements between each other regarding projects that affect us. In comparison with the two other previous levels of priority that required a global scale of judgement for the general prohibition, and a local scale of judgement for the development of nurturing milieus, this latter area of ethics covers the small scale of interactions between agents.

3.3.3 Transmission and the phenomenological agent

What does the working definition of sustainability and its normative implications mean for the phenomenological agent? The setting of priorities and the evaluation of projects and worldviews are to be done from the perspective of the observer. The assessment of projects regarding their viability related to the general prohibition requires scientific knowledge and communications between assessors who attempt to take a neutral standpoint and conduct a reasoning as much as possible objective. Nevertheless, most of the imprints we contribute to leaving on the milieu are coming from our behaviours in daily life, in which we are irreducibly trapped in the perspective of the phenomenological agents, with our blind spots and a limited time for reflection. Most of us have only few opportunities to take a step back and evaluate from a tentative observer perspective the situation and the projects that we witness or in which we engage. Then, we are as individual agents the last resort for ethical decision-making, while we are at the same time glued to our lifestyles. In other words, only individual agents can make sense of the normative implications described here and apply them to their own actions in order to bring a change in the world. But simultaneously, we are ourselves constantly in cycles of codetermination with our milieus, in which immediate demands for action and attention keep distracting us from our ethical drive.

From the perspective of the individual phenomenological agent, the definition of sustainability seems overly abstract. The "maintenance of the conditions of possibility of continuation of self-determining human existences" appears quite distant from the concerns of everyday life. A reason for that is that the individual human agent is going to die sooner than later. Why would one care for the maintenance of the possibility of continuation for human existences to the point of making changes in her own present lifestyle, given that her own individual and precious life is already so short? An element to reply to this question can be found in the cycles of codetermination between the self and the milieu. The individual cannot be extracted from her milieu without enduring deep changes in herself. In other words, her own milieu – or at least some elements of it – is a constitutive part of who she is.

The individual agent will inevitably die, whereas her milieu is likely to survive her and nurture and shape over human beings after the individual's death. Then, from the individual perspective, more than "maintaining" her environment and milieu, she is contributing to transmit them to others. By definition, transmission is the process of passing something from one person to another. Even when the agency of the individual stops with her death, her imprints on the milieu keep having impacts. The same goes when the individual definitely leaves a particular milieu. Her impacts do not stop with her absence. Meanwhile, other individual agents will be taking the milieus she left in their own hands. Because of our unescapable individual death, sustainability is not only the maintenance, but also the *transmission* of a living meaningful milieu and of a healthy environment.

My working definition of sustainability then amounts to keep going the cycles between human beings and their milieu. For the phenomenological agent, it means evaluating her own actions and lifestyles relatively to their imprint on the milieu, that is, relatively to how they contribute to shaping the milieu as a matrix for other human beings. Improving the matrix becomes the normative direction of ethical actions. "Improving" here refers to changing the matrix according to the normative implications discussed before. This process of improving the matrix will, in the long run, benefit future generations. But before then, it will also likely benefit the individual herself, because she will enjoy more meaning, peace and safety as she goes through the vulnerable stages of her own life, towards old age and death.

We all care deeply for some aspects of the world, be it some values and worldviews we hold dear, or some projects we dedicated our life to make successful. The prospect of the continuation of these values and projects contributes greatly to our subjective sense of well-being and happiness. Our imprints on the milieu are decisive in the future flourishing of these values and projects, or in their jeopardy. Values, ideals and projects thus exist in the milieu, which, in turn, as a matrix, shapes other human individuals, and inspire them to take over. Then, the specific characteristic of future generations is less a concern to us than designing meaningful and nurturing milieus that will shape them. In other words, we are constantly connected with future (and past) generations through our milieu as an imprint and as a matrix. Focusing on the transmission of a meaningful milieu can provide justifications for caring for future generations (Droz 2019b). In short, we might not care abstractly for the maintenance of the possibility of continuation for human existences, but we definitely care for the prospects of the main projects and ideals we dedicated our lives to, from our children and our community to our posthumous reputation. What precisely this project is differs for each of us, but the possibility of its continuation requires us to make changes in our present lifestyles, as they are pieces of our milieu, pieces whose existence depends directly on the intertwinement of webs of relations of our milieu as a whole (and possibly of the multiple interconnected milieus across the globe).

Last but not least, the importance of self-determination is highlighted when we take the perspective of the phenomenological agent. I described how we are shaped and influenced by our milieu as a matrix, through participatory sense-making, the cultural imaginary and practices. But I also insisted on the fact that we are *co*determining ourselves with the milieu, that is, we are not solely passively determined by it, but we still, always, have a window of choices to exercise our own agency. To have this window of choice is a necessary condition for ethics. If individuals were completely constrained and determined by their milieu, then there would be no hope for ethics and the whole discussion of this book would become void. As I described when I discussed individual imprints, our ethical agency appears in every of our actions and omissions. Every of our choices leads us to become who we are, to gain the habits and automatisms that we have and to see the world and act on it in the way we do now. In other words, our ethical agency lies in every choice we make, more or less deliberately, at every breath we take.

To erect self-determination as a normatively desirable outcome amounts to argue for enlarging the window of choices that we have, with the hope that increasing our range of possible choices of actions will enable us to take better actions and improve our imprint on the world. To use the metaphor of the dynamic interlocking between the self and the milieu, we need to temporarily take our distances and increase the space in-between our self and our milieu in order to critically think and take ethical decisions. Ethics emerges from this interval in-between, and can develop and flourish only through this in-between space. Without this agency and the possibility to act otherwise, there would be no place for ethics. We need self-determination for ethical responsibility.

3.4 Summary

The working definition of sustainability intertwines the ideas of precaution, diversity and autonomy: Sustainability is the maintenance of the conditions of possibility of continuation of (1) self-determining flourishing human existences. It entails (2) maintaining the general processes of the global environment autonomous and healthy to limit the possible harmful consequences of the conflicts of distribution and domination, and (3) cultivating meaningful, diverse and adaptable nurturing milieus. To do so, we need to minimize all types of harm, not only harm affecting human health and human quality of life, but also harm that affects mainly the nonhuman natural environment. Preserving a healthy global natural environment is crucial especially because it can neutrally provide healthy living conditions for human communities without forcing them to rely on highly developed technologies whose development, knowledge and access is likely to be restricted to dominant groups. Nevertheless, the protection of the autonomy of the global environmental system does not protect us from other conflicts between human groups and within groups. These conflicts are even more complex as patterns of distribution of resources and power, and dynamics of oppression and domination are often central and meaningful characteristics of the social structure of a particular milieu. Finally, the realization of what appears to the individual agent as a flourishing life might directly conflict with the preservation of the environment, and with the maintenance of meaningful milieus, especially if this image of a flourishing life rests on relational dynamics of domination and power over other human beings. Self-determination must take place within the limitations of some normative safeguards, and this is the object of Chapter 4.

Table 3.2 synthesizes some normative implications that were drawn from the working definition of sustainability. Finally, I pointed out some limits and priorities regarding the application of the normative implications of the working definition of sustainability. While I placed self-determination as desirable in my working definition of sustainability, some might reject individual self-determination as a normative value. I discussed this objection and presented some limited ways to bend my argumentation to call to proponents of systems rejecting individual self-determination (such as worldviews valuing more heresy and apostasy than self-determination). This brought me to propose some priorities in the evaluation of projects and worldviews. The first highest priority is the respect of the necessary and precautionary prohibition. Then comes the criteria of no harm, and finally the criteria of continuation. Finally, from the perspective of the phenomenological agent, my working definition of sustainability implies the idea of transmission of meaningful milieus, including the projects and values that we hold dear during our life. I also showed that from the perspective of the phenomenological agent, self-determination is essential as it is the place for agency, that is, a necessary space for ethics.

Table 3.2 Normative implications of sustainability

Working definition of sustainability

Sustainability is the maintenance of the conditions of possibility of continuation of (1) self-determining flourishing human existences. It entails (2) maintaining the general processes of the global environment autonomous and healthy to limit the possible harmful consequences of the conflicts of distribution and domination, and (3) cultivating meaningful, diverse and adaptable nurturing milieus.

Criteria of evaluation of projects, practices and lifestyles

1   *Harm*: Minimizing harm, including harm affecting the health and quality of life of other human beings, mainly nonhuman environmental elements, and indirect harm affecting elements of the milieu.
2   *Continuation*: Fostering the possibility of continuation of the practice in the long term.

Necessary and precautionary prohibition

Human beings have, individually and collectively, a necessary prohibition from depleting or destroying the global environmental systems to the point where these systems cannot independently provide healthy living conditions to current and future human beings.

Virtues of holistic conduct

* *Curious*: Seek encounter and exchange with diverse unknown others.
* *Adaptable*: Be ready to change our lifestyle, habits and (part of) our worldview, namely, avoid dogmatic walls that prevent conscious change and exclude some experiences (including thought and conceptual experiences).
* *Humble*: Limit caring by understanding the irreducible uncertainty relative to the other's needs.
* *Self-reflective*: Carefully choose in what projects and worldviews we invest our emotions, beliefs and trust.

Note

1   Some aspects presented in this chapter were already mentioned in Droz (2019a, 2019b).

Bibliography

Alwi, Engku A. Z. E. et al. 2015. 'Heresy in Malaysia: An Analysis | Mediterranean Journal of Social Sciences'. *Mediterranean Journal of Social Sciences*, 6 (2): 467–68. https://www.richtmann.org/journal/index.php/mjss/article/view/5925

Barry, Brian. 1999. 'Sustainability and Intergenerational Justice'. In *Fairness and Futurity: Essays on Sustainability and Justice*. Oxford: Oxford University Press. https://oxford.universitypressscholarship.com/view/10.1093/0198294891.001.0001/acprof-9780198294894-chapter-5.

Beckerman, Wilfred. 1999. 'Sustainable Development and Our Obligations to Future Generations'. In *Fairness and Futurity: Essays on Sustainability and Justice*, edited by Andrew Dobson, 71–92. Oxford: Oxford University Press.

Bonica, John. 1979. 'The Need of a Taxonomy.' *Pain* 6 (3): 247–48. https://doi.org/10.1016/0304-3959(79)90046-0.

Bruggen, Ariena H. C. van, Erica M. Goss, Arie Havelaar, Anne D. van Diepeningen, Maria R. Finckh, and J. Glenn Morris. 2019. 'One Health - Cycling of Diverse Microbial Communities as a Connecting Force for Soil, Plant, Animal, Human and Ecosystem Health'. *Science of the Total Environment* 664 (May): 927–37. https://doi.org/10.1016/j.scitotenv.2019.02.091.

Christman, John. 2018. 'Autonomy in Moral and Political Philosophy'. *2018, "Autonomy in Moral and Political Philosophy" The Stanford Encyclopaedia of Philosophy*. https://stanford.library.sydney.edu.au/archives/spr2008/entries/autonomy-moral/.

Droz, Laÿna. 2019a. 'Redefining Sustainability: From Self-Determination to Environmental Autonomy'. *Philosophies* 4 (3): 42. https://doi.org/10.3390/philosophies4030042.

———. 2019b, November 18. 'Tetsuro Watsuji's Milieu and Intergenerational Environmental Ethics'. *Environmental Ethics*. https://doi.org/10.5840/enviroethics20194114.

Escobar, Arturo. 1995. *Encountering Development: The Making and Unmaking of the Third World*. STU-Student edition. Princeton University Press. https://www.jstor.org/stable/j.ctt7rtgw.

Faith, Daniel P. 2019, Fall. 'Biodiversity'. In *The Stanford Encyclopedia of Philosophy*, edited by Edward N. Zalta. New York: Metaphysics Research Lab, Stanford University. https://plato.stanford.edu/archives/fall2019/entries/biodiversity/.

Goodin, Robert. 1992. *Green Political Theory*. Cambridge; Cambridge, MA: Polity Press.

Harris, Graham. 2007. *Seeking Sustainability in an Age of Complexity*. Cambridge University Press.

Holland, Alan. 1999. 'Sustainability: Should We Start From Here?' In *Fairness and Futurity: Essays on Environmental Sustainability and Social Justice*, edited by Andrew Dobson. Oxford: Oxford University Press.

IPBES. 2018. The IPBES Assessment Report on Land Degradation and Restoration. In *Secretariat of the Intergovernmental Science-Policy Platform on Biodiversity and Ecosystem Services*, edited by. Montanarella, Luca, Scholes, Robert, and Brainich, Anastasia, Bonn, Germany, 744 pages. https://doi.org/10.5281/zenodo.3237392

———. 2019. The Global Assessment Report on Biodiversity and Ecosystems Services. In *Intergovernmental Science-Policy Platform on Biodiversity and Ecosystem Services*, edited by Eduardo S. Brondizio, Josef Settele, Sandra Díaz, and Hien Thu Ngo. Bonn: IPBES Secretariat, UNEP.

———. 2020. IPBES Workshop on Biodiversity and Pandemics. In *Intergovernmental Platform on Biodiversity and Ecosystem Services*, edited by Peter Daszak, John Amuasi, Peter Buss, Carlos Das Neves, Heliana Dundarova, Yasha Feferholtz, Gabor Foldvari, David Hayman, Etienosa Igbinosa, Sandra Junglen, Thijs Kuiken, Qiyong Liu, Benjamin Roche, Gerardo Suzan, Marcela Uhart, Chadia Wannous, Katie Woolaston, Carlos Zambrana Torrelio, Nichole Barger, David Cooper, Tom De Meulenaer, Hans-Otto Poertner, Cristina Romanelli, Karen O'Brien, Paola Mosig Reidl, Unai Pascual, Peter Stoett, Hien Thu Ngo. Bonn: IPBES secretariat. doi:10.5281/zenodo.4147317

IPCC, 2019: Summary for Policymakers. In: Climate Change and Land: an IPCC special report on climate change, desertification, land degradation, sustainable land management, food security, and greenhouse gas fluxes in terrestrial ecosystems, edited by Valérie Masson-Delmotte, Panmao Zhai, Hans-Otto Pörtner,

Debra Roberts, Jim Skea, Eduardo Calvo Buendía, Priyadarshi R. Shukla, Raphael Slade, Sarah Connors, Renée van Diemen, Marion Ferrat, Eamon Haughey, Sigourney Luz, Suvadip Neogi, Minal Pathak, Jan Petzold, Joana Portugal Pereira, Purvi Vyas, Elizabeth Huntley, Katie Kissick, Malek Belkacemi, Juliette Malley. In press.

Jacques, Peter. 2020. *Sustainability: The Basics*. Abingdon: Routledge.

Jamieson, Dale. 2008. *Ethics and the Environment: An Introduction*. Cambridge Applied Ethics. Cambridge: Cambridge University Press. https://doi.org/10.1017/CBO9780511806186.

Kagan, Sacha. 2010. 'Cultures of Sustainability and the Aesthetics of the Pattern That Connects'. *Futures*, Global Mindset Change, 42 (10): 1094–101. https://doi.org/10.1016/j.futures.2010.08.009.

Kymlicka, Will. 1991. *Liberalism, Community, and Culture*. Oxford: Oxford University Press.

———. 1996. *Multicultural Citizenship: A Liberal Theory of Minority Rights. Multicultural Citizenship*. Oxford University Press. https://oxford.universitypressscholarship.com/view/10.1093/0198290918.001.0001/acprof-9780198290919.

Latouche, Serge, and Simon Leung. 1996. *The Westernization of the World: Significance, Scope and Limits of the Drive towards Global Uniformity*. Cambridge; Malden, MA: Polity Press.

MacIntyre, Alasdair. 2007. *After Virtue: A Study in Moral Theory*, 3rd Edition. Notre Dame, IN: University of Notre Dame Press.

Macioce, Fabio. 2012. 'What Can We Do? A Philosophical Analysis of Individual Self-Determination'. *Eidos* 16 (January): 100–29.

Mackenzie, Catriona, and Natalie Stoljar. 2000. *Relational Autonomy: Feminist Perspectives on Autonomy, Agency, and the Social Self*. Oxford: Oxford University Press.

Mestre, Mireia, and Juan Höfer. 2020, December. 'The Microbial Conveyor Belt: Connecting the Globe through Dispersion and Dormancy'. *Trends in Microbiology*. https://doi.org/10.1016/j.tim.2020.10.007.

Mühlhäusler, Peter. 1995. 'The Interdependence of Linguistics and Biological Diversity'. In Northern Territory University Press. https://digital.library.adelaide.edu.au/dspace/handle/2440/30170.

Müller, Frank G. 2013. 'The Discounting Confusion: An Ecological Economics Perspective'. *Economia* 36 (71): 57–74.

Norris, Fran H., Susan P. Stevens, Betty Pfefferbaum, Karen F. Wyche, and Rose L. Pfefferbaum. 2008. 'Community Resilience as a Metaphor, Theory, Set of Capacities, and Strategy for Disaster Readiness'. *American Journal of Community Psychology* 41 (1–2): 127–50. https://doi.org/10.1007/s10464-007-9156-6.

Ostrom, Elinor. 2015. *Governing the Commons, The Evolution of Institutions for Collective Action*. Cambridge: Cambridge University Press.

Pascual, Unai and Eneko Garmendia. 2013. 'A Justice Critique of Environmental Valuation for Ecosystem Governance. The Justices and Injustices of Ecosystem Services'. In *The Justices and Injustices of Ecosystem Services* edited by Thomas Sirkor, 161–86. New York: Routledge. https://doi.org/10.4324/9780203395288-25.

Rahnema, Majid, and Victoria Bawtree. 1997. *The Post-Development Reader*. London; Atlantic Highlands, NJ: Dhaka; Halifax: Cape Town: Zed Books Ltd.

Rock, Melanie J., and Chris Degeling. 2015. 'Public Health Ethics and More-than-Human Solidarity'. *Social Science & Medicine*, One World One Health? Social

Science Engagements with the One Medicine Agenda, 129 (March): 61–67. https://doi.org/10.1016/j.socscimed.2014.05.050.

Roger Few. 2013. 'Health, Environment and Ecosystem Services Framework: A Justice Critique'. In *The Justice and Injustice of Ecosystem Services*, edited by Thomas Siker, 140–60. Abingdon: Routledge.

Rozzi, Ricardo. 2015. 'Implications of the Biocultural Ethic for Earth Stewardship'. In *Earth Stewardship, Ecology and Ethics 2* Ricardo Rozzi (ed.), Cham: Springer International, 113–36. Switzerland.

Rozzi, Ricardo. 2018. 'Biocultural Homogenization: A Wicked Problem in the Anthropocene'. In *From Biocultural Homogenization to Biocultural Conservation, Ecology and Ethics*, edited by Ricardo Rozzi, Roy H. May Jr., F. Stuart Chapin III, Francisca Massardo, Michael C. Gavin. Irene J. Klaver. Anibal Pauchard, Martin A. Nunez, Daniel Simberloff, 21–47. Cham: Springer Nature.

Sarkar, Sahotra. 2010. 'Diversity: A Philosophical Perspective'. *Diversity* 2 (1): 127–41. https://doi.org/10.3390/d2010127.

Sidgwick, Henry. 1874. *The Methods of Ethics*. London, MacMillan and Co. (7th Hackett reprint), ISBN 0-915145-28-6. https://en.wikipedia.org/wiki/The_Methods_of_Ethics

Skutnabb-Kangas, Tove, Luisa Maffi, and David Harmon. 2003. *Sharing a World of Difference. The Earth's Linguistic, Cultural, and Biological Diversity*. Paris: UNESCO. https://unesdoc.unesco.org/ark:/48223/pf0000132384

Stelios, Virvidakis. 2014, October 1. 'Living Well and Having a Good Life: Interpreting the Distinction'. *Philosophical Inquiry*. https://doi.org/10.5840/philinquiry2014383/411.

UNEP and ILRI. 2020. 'Preventing the Next Pandemic: Zoonotic Diseases and How to Break the Chain of Transmission'. Nairobi, Kenya. https://www.unep.org/resources/report/preventing-future-zoonotic-disease-outbreaks-protecting-environment-animals-and

Verweij, Marcel, and Bernice Bovenkerk. 2016. 'Ethical Promises and Pitfalls of OneHealth'. *Public Health Ethics* 9 (1): 1–4. https://doi.org/10.1093/phe/phw003.

Wilson, Geoff A. 2018. '"Constructive Tensions" in Resilience Research: Critical Reflections from a Human Geography Perspective'. *The Geographical Journal* 184 (1): 89–99. https://doi.org/10.1111/geoj.12232.

World Resources Institute. 2005. *Millennium Ecosystem Assessment, Ecosystems and Human Well-Being: Biodiversity Synthesis*. Washington, DC: Synthesis. Island Press. https://www.millenniumassessment.org/documents/document.356.aspx.pdf

4 Responsibility

Individual responsibility regarding sustainability is the cornerstone of the motivational framework that supports sustainable behaviours that this book aims at developing from the concept of milieu.[1] This motivational framework aims towards sustainability, as shown in the upper part of Figure 4.1, which synthesizes it. The three-level model (Figure 4.1) that encompasses the individual, the milieu and the spatiotemporal environment is represented on the background in concentric areas enveloping the phenomenological agent. The intermediary level of the milieu bridges the neutral spatiotemporal environment and the phenomenological agent. In the middle, the framework of the milieu bounds together the phenomenological agent, the medial imprint, the community and the medial matrix in dynamic cycles. Finally, the individual phenomenological agent faces ethical choices and must balance her agency and vulnerabilities, which echoes the balance of contributory responsibility and capacity responsibility.

What is individual *moral* responsibility for environmental harm within the framework of the milieu? Responsibility is the duty to answer for one's actions and the consequences caused by them, that is, to assume them and, in case of harm, to take reparative actions. There are three dimensions in this concise definition. First, responsibility has to do with the consequences of one's action. It raises the question of what counts as causation and causal contribution. This question is challenging, especially if we take into consideration the multiple scales of consequences, including mediated and domino-effect consequences as discussed in Chapter 2. The second dimension has to do with the readiness to be answerable or to be held accountable for something. It relates to the capacity of the agent and touches aspects of mental health and freedom. The third dimension addresses the question of what an agent is required to do when it is established that she caused a specific harmful consequence, and that she is accountable for it. Without this more concrete aspect of what reparative actions an agent recognized as responsible for an action must do, responsibility would remain a floating concept void of useful applications. In the literature, this third dimension often appears under reflections about punishment.

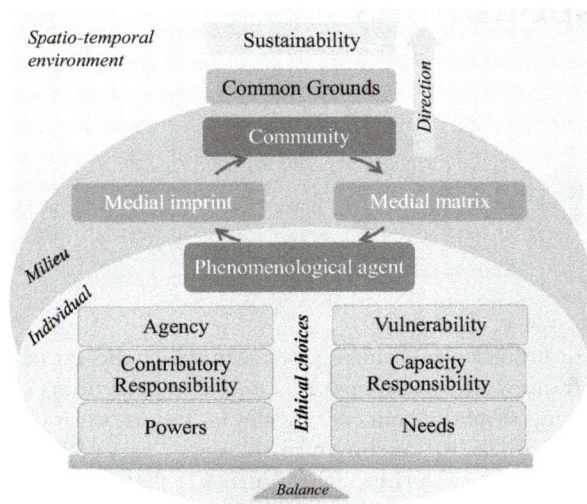

Figure 4.1 Motivational framework of the milieu, including responsibility for environmental harm, sustainability as a normative direction and the three-level model

Understood in these terms, responsibility appears to connect a past action with a future expected action. Responsibility can be used as a hinge between what the agent did in the past and what she is ethically required to do in the future. From this perspective, the two dimensions of causal contribution and answerability are more retrospective dimensions, while the dimension of reparative actions is prospective. This last dimension requires further investigations into what is regarded as good, and about towards what ideals the reparative actions taken should be oriented. In Chapter 3, I suggested that the idea of sustainability provides such an ideal of good that can be compatible with different worldviews.

Up to now, I argued that an individual is simultaneously shaped by the milieu and shaping her milieu. The individual is shaped by the milieu as a matrix because she is making sense of the world in a relational and participatory way, she is borrowing concepts and ideals from the cultural imaginary and she acts guided and constrained by practices rooted in the milieu. The medial matrix shapes the capacity of the individual, namely, if she thinks of herself as a potentially responsible agent or not, what she thinks is good, what she thinks she can and she is required to do by the social expectations surrounding her, etc. Of course, these influences do not eliminate the agent's capacity to reflect, think and make decisions that might run counter to socially expected practices and worldviews. But they strongly contribute to shaping the individual reasoning and range of possible decisions and actions.

The individual also leaves imprints in the milieu. These imprints can be direct, that is, all the significant traces left by the action of an individual. They can also have domino-effect consequences spanning up to the global level. And they can also be mediated, propagated, multiplied and sometimes orchestrated by the actions of other agents who are members of the same social structure. If, as suggested by the idea of medial imprint, we include all these consequences into the realm of what an agent is responsible to have caused, we run into a *reductio ad absurdum* objection. Indeed, one could appeal to extremes and argue that an agent is then somehow contributorily responsible for everything happening in the world. As a result, the very idea of responsibility would become absurd, as it amounts to saying that everybody is responsible for everything, and so eliminating the ethically pressuring force of the idea of responsibility. This argumentation can be heard in laymen discussions and in some political debates about the assignment of responsibility for environmental problems such as climate change. Notably, the wide-encompassing account of imprint on which my account of moral responsibility is based includes far-reaching consequences that might be excluded under the common understanding of causation. For this reason, instead of "causal responsibility", I carefully prefer to speak of "contributory responsibility" as it encompasses consequences that the agent contributed to cause, but which she is not the sole contributor. In some cases, her contribution alone might even have made no significant difference in the occurrence of the consequence.

Three lines of answers can be developed. First, one can object the extrapolation because any social structure has some limits as to who and what is included and what is excluded. Similarly, any domino-effect consequence is also limited in scope, in the sense that the effects in spatiotemporal realms far distant from the action are insignificant. Thus, it is not plausible to argue that an agent is responsible for everything happening in the world.

Second, one can define the concept of responsibility in a way that does not restrain it to causal responsibility. Indeed, the limitless extrapolation of the contributory responsibilities of an agent needs to be blocked by the limited scope of what an agent can actually do. In other words, the risk of *reductio ad absurdum* in the assignment of responsibilities is to be nuanced, because responsibility must be balanced with the agent's mental health. To knock down an individual agent with guilt related to overwhelming responsibilities for the role she played in the occurrence of harmful events is not desirable. In our contemporary world, many feel hostage of a harmful system in which they have little to say. Regardless of the question if this claimed powerlessness is a convenient excuse or a real limitation, an agent considering herself powerless and experiencing overwhelming guilt and pain is unlikely to take reparative actions. On the contrary, she is likely to turn away from the perceived source of these unpleasant feelings, putting into doubt the existence of the problem (e.g. climate change denying), or playing down her personal responsibility in its occurrence (e.g. powerless hostage of states and corporations). From a pragmatist standpoint, contributory responsibility thus needs to be balanced with the individual's capacity to be held responsible.

Third, responsibility can be argued to be non-binary. It can be derived in a wide range of degrees and forms, each requiring different kinds of reparative actions and reactive attitudes. The farmer spreading herbicide on her field could be held fully responsible for the local loss of biodiversity in the surroundings of her own field, while being held only partially responsible for the contamination of the underground waters of the neighbouring village. This is, of course, debatable, and further investigation is needed to clarify what kind of reparative actions are required for what kind of harmful consequences.

In this chapter, I develop an account of individual moral responsibility for environmental problems weaving around these three lines of arguments. I first explore what counts as contributory individual responsibility in the case of environmental problems. Then, I clarify the capacity responsibility that an agent has by virtue of being a member of a social structure. Finally, I discuss what reparative actions are expected by the assignment of moral responsibility.

4.1 Imprints and contributory individual responsibility

A fundamental dimension of responsibility is intrinsically linked to a relation of causation between the agent and an event. A case of contributory individual responsibility is when responsibility is assigned for and limited to the consequences that are caused directly by the action of an agent. It can also include omissions, in other words, inaction. Debates in philosophy and in law about what consequences are considered to be caused by an agent's action (or inaction) are ubiquitous and have been going on in different traditions for thousands of years. Instead of diving into these details, for the purpose of discussing responsibility for environmental problems, I accept a wide-encompassing definition of causal contribution. I assume that it is enough for the agent to have made a difference in the occurrence of an event in the world for her to be considered to have played a role in causing this event. Conversely, if the agent had not supported a certain state of affairs, then the event would not have occurred. This includes supporting a state of affairs by inaction and by refraining to take actions to prevent its continuation and to protest against it.

This wide-encompassing account of causal contribution is clearly not equivalent to individual moral responsibility as a whole. If so, it would frontally collide with a widely shared intuition related to questions of freedom and control of the agency. For instance, Aristotle already argued that legal responsibility could be assigned only if the action was voluntary, namely, if the agent had the capacity to act otherwise (Golding 2005; Aristotle 2014). According to him, an action is involuntary when the agent is coerced or ignorant of relevant elements. Coercion can be internal (e.g. mental limitations, drugs) or external (e.g. if the agent's hands are chained). Ignorance of laws and relevant facts also affects the voluntariness of an action. When it comes to assigning contributory responsibility to an individual agent for her action and its consequences, it is hard to separate what pertains uniquely to causal contribution, and what involves judgements of capacity. So, from the perspective of the observer, even though a causal relation was established between an agent and an event, it is not enough to hold the agent responsible for this event. In order to clarify what the criteria appropriate to assign contributory responsibility are, it is first necessary to explore the relationship between actions, events and consequences. I will do so in connection with medial imprints.

4.1.1 Actions, consequences and imprints

At the very beginning of this chapter, I suggested that responsibility connects a past action with a future expected action. What is this connection and what are the roles of causal contribution and ethics in this picture? Let's start with the most basic case of an agent doing a specific action provoking an externally observable event. The English word "agency" is almost as useful as it is unclear. It can refer to an action or an intervention producing a particular effect, and to the thing or person (e.g. the agent) that acts to produce a particular result. In philosophy, the question of the nature of actions deserves a whole field named philosophy of action. To keep it simple, let us understand actions as bodily movements that can be associated with mental activities (e.g. Davidson 1980). This gives us already the twofold aspects of action. From the perspective of the phenomenological agent, actions are often recognized as intentional, sometimes including automatic actions. The easiest example of an action is then the bodily movement of an agent done intentionally for a specific purpose (Figure 4.2). From the perspective of the observer, actions can be seen as events and instances of a causal relation. What counts as the action, what counts as the event and what counts as a consequence depend then on the theory of action that is endorsed.

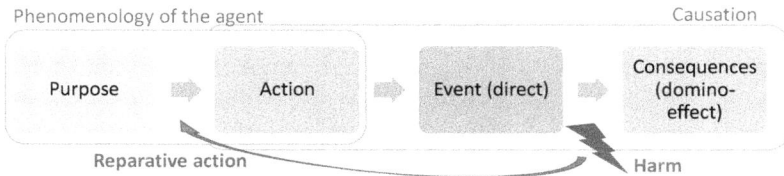

Figure 4.2 Responsibility connects a past action with a future expected reparative action

As sketched in Figure 4.2, from the perspective of the agent, she is doing an action, sometimes for a specific purpose, sometimes without thinking about it (automatically), and this action contributed to producing a direct consequence in the world, that is, an event. The agent might not be aware of this direct event. Moreover, she might be completely blinded to the domino-effect consequences. From the perspective of the observer, we can see an action as a bodily movement, which causes an event as a direct consequence, and possible domino-effect consequences. As I suggested before, if the event is causing harm, then the agents responsible for the action producing this event may be required to take reparative action.

What makes an agent the potentially responsible owner of what seems to be her action? An agent might be pushed to take an action by irrepressible urge and desires, or she might be doing it without clear purpose, or even completely automatically, without being aware of it. In short, the agent might not have been free to do otherwise. If we attempt to find a contributory relation sufficient to assess responsibility in the action-event pair, excluding the circumstances and past choices that brought the agent to this situation, then we are confronted with the mysterious black box of the agent's phenomenological experience. Fischer and Ravizza propose to bypass this obstacle by situating the key criteria for responsibility in guidance control, regardless of the accessibility of alternative (Fischer and Ravizza 1998). According to them, an agent has guidance control if she recognizes herself as the source of the action, namely, if she owns the mechanism doing the action (usually, the body), and as a relevant candidate for social reactions. This latter requirement is inspired by Strawson who suggested that an agent is responsible "insofar as she is an appropriate candidate for the reactive attitudes" (Strawson 1962). For Fischer and Ravizza, it is enough that the agent has, in the past, taken responsibility as an agent for her to be considered responsible for her later actions and omissions. Usually, they argue that individuals make certain kinds of mechanisms their own by taking responsibility for them partly as a result of education. In the early childhood, the individual learns to see upshots in the world depend on her own choices and bodily movements, and in turn, to see herself as a fair target of reactive attitudes by others, such as punishment and praise (Fischer and Ravizza 1998, 241). Through this historical process of seeing themselves as agents, individuals are taking responsibility for their future actions and omissions.

This view situating the key criteria for responsibility in guidance control escapes the traps of the freedom-determinism debates and the black box of the phenomenological state of the agent at the moment of taking an action. Moreover, it gives symmetrical treatments to actions and omissions as instances of guidance control. This meets our need for a criterion for responsibility as strong for omissions and inactions as for actions, as most of the environmental problems unfold as much because of the inaction of most individuals, as because of the actions of few. Finally, it recognizes responsibility as the result of an historical process inseparable from the web of social reactions.

When it comes to environmental problems, it is often impossible to isolate one single action from its temporal and social circumstances and to pinpoint it as the sole cause of a harmful event. In contrast with the action of pulling the trigger to kill someone, the action of spreading herbicide on one's own field produces severe, harmful and irremediable consequences only (or mainly) when it is repeated numerous times over the course of years. Most single actions and their related consequences are closely entangled with ways of life and projects rooted in the past of the individual, inspired by the milieu and the social structure and spanning often to the future of the individual. So, we need a criterion to assign contributory responsibility that can encompass harmful consequences emerging from repeated actions. This is not new. A spouse diluting small amounts of poisons in the food of her partner over years, leading to the death of the latter, is still usually considered guilty for murder. Spreading herbicide and improperly disposing of wastes can be considered in the same way.

Actions are also often glued to lifestyles. Driving long distances every day in a private car to commute to work has detrimental effects on air pollution and climate change (ignoring for now the problem of cumulative effect discussed later). But even if we take the sum of all the rides over one or several years, it does not give us the whole story. Indeed, the driver's decision to commute in individual car is intrinsically linked with her choice of lifestyle, where she lives, what work she does, etc. As we have seen previously, it is also intertwined with her own ideals about what her lifestyle should be, guided by the social expectations surrounding her, and the design of the social structure and milieu. The fact that the agent is today too driving long distances to her work and contributing to severe air pollution is not the result of her decision in the morning, but of all the choices and actions she made over her life, guided and constrained by her milieu.

The driver might argue that she cannot do otherwise because there is no public transportation, that housing closer to her workplace is not affordable, etc. In short, she could argue that she is "hostage of the system", ranging from social expectation requiring her to have a private car and to keep a specific standard of living, to material constraints such as the absence of supportive mechanisms allowing her to avoid using her private car daily. In other words, the agent could reject responsibility for the state of affairs that constrains her to lead such a lifestyle. She could use her passivity as an excuse to reject responsibility for a state of affairs. But under an account of contributory responsibility treating omissions and inaction symmetrically to actions, then she would also be responsible for not doing anything to prevent a state of affairs to be established, or to change the current state of affairs.

These cases of consequences produced by repeated actions, lifestyles and passivity towards state of affairs urge us to zoom out from the micro level of the action–event pair. The idea of imprint as developed in Chapter 2 gives us much more inclusive glasses to look at responsibility for environmental problems. I defined previously medial imprints as the traces left on the milieu by the actions of human agents, and I later extended this to include inaction and mediated imprints coming from the very existence of an individual because of the interconnectivity of the social structure. Imprints thus include the consequences on the milieu coming from isolated actions, repeated actions, inaction, lifestyles and passivity towards a state of affairs. The central question of contributory responsibility is then turned upside down. It no longer concerns what the agent contributed to cause in the world, and if she has done so voluntarily or not. The main question becomes what the agent could and should have done to prevent the occurrence of a harmful event or state of affairs. In other words, what the individual agent should do to leave less harmful or more positive imprints on her milieu, assessed as harmful or positive regarding sustainability as discussed in Chapter 3.

The question "what the agent should do" is not strictly restricted to a specific situation where she did or refrained from doing a specific series of bodily movements, but extends to questions of choices of lifestyles, choices of worldviews, choices of values, etc. Responsibility is not confined to a specific action at one specific temporal point, but it expands to the whole individual life and evolution that brought her to being in this specific situation and that brought her to see this specific situation from a certain perspective. In short, we are not only responsible for what we do, but, more importantly, we are responsible for what we become. This expansion of responsibility allows us to deal with automatisms and habits. It allows us to treat the ubiquitous cases where the agent is not seeing other alternative behaviours because she is not questioning herself, her automatisms, her habits and her worldview. Indeed, under such an encompassing account of responsibility, the agent is then – partially – responsible for holding some beliefs and worldviews leading her to see and act in the world in a certain way and restricting the range of her possible actions.

Whatever she does or refrains from doing, the agent is leaving imprints in the world. Moreover, because the agent is (or at the very least has been, during infancy) inseparably entangled with the social structure she is a member of, her very existence is leaving imprints on the milieu, imprints that are mediated by other people's actions. Then, the fact that the agent has at least partly caused or contributed to causing the direct, mediated and domino-effect imprints is assumed. Furthermore, to apply Fischer and Ravizza's guidance control criteria, if the agent recognizes herself as an agent capable of doing her own actions and deserving reactive attitudes of praise and blame for them, she must then be held somehow responsible for her medial imprints. In other words, the agent cannot claim ownership for some of her imprints and reject it for others. This already brings us to a first conclusion that prior to questions of free will and agency, because of human fundamental dependency to the social nets, every individual has an irrevocable contributory responsibility for her imprints, direct and mediated. This claim echoes Hans Jonas' idea of natural responsibility that is irrevocable and usually attached to an existing being, the archetype being parental responsibility towards children (Jonas 1979).

To argue for irrevocable contributory responsibility does not eliminate questions of degrees of responsibility, and of what kind of reparative actions it entails. It is crucial here to note that by claiming that we all have an irrevocable contributory responsibility for our imprints, I do not claim that we all have an *equal* moral responsibility regardless of the accessibility of alternative of behaviours. Similarly, I do not claim that we are all required to undertake the *same* reparative actions. To address these aspects, we need to investigate the capacity of the agent to be answerable, related to practical resources and mental health. And we need to assess what the agent can do as reparative action, which will depend on the agent's contribution and capacity. Up to now, I have barely argued for a wide-encompassing conception of contributory responsibility as a reflection of the idea of individual medial imprint.

4.1.2 Source of harm, cumulative effects and collective imprint

In the account of causal contribution I defend here, to be considered contributory responsible, it is enough for the agent to have made a difference in the occurrence (including the continuation) of an event or a state of affairs. This includes letting something happen, namely, passive support. And it also covers giving advice to another agent, leading the latter to act in certain ways. In these cases, contributory responsibility is mediated by inaction and actions of other agents influenced and encouraged by the first agent. It is still possible to claim that without the individual agent's action or omission, a specific event would not have occurred, or a specific state of affairs would not continue. But there are many environmental problems for which it is hard to draw such a causal line between the individual agent and a global harmful effect. Indeed, climate change is the archetype of causation determined by cumulative effects. The single action of burning charcoal to cook a family meal cannot be considered solely contributing causing climate change. Even the specific local lifestyle based on charcoal cooking, alone, can hardly be considered responsible.

Examples of imperceptible but cumulative impacts producing severe harm have become worryingly common. Beyond environmental problems, online shaming and hate speech on the Internet are often presented as cases in which no individual or relatively small group of individuals can be pointed as the culprits. In these cases, responsibility seems to be diluted in the masses, between tens of thousands of individuals around the globe who, from their own perspective, have only done a seemingly insignificant action such as sharing a specific social media content, or spreading herbicide. The individual agent writing a hateful comment is likely to recognize herself as responsible for the action of posting the comment, but not for the cumulative consequence of the suicide of the teenager to whom this comment was addressed, brought about by thousands of similar comments written by multiple other individuals. The writer of the comment then is likely to not experience tremendous feelings of guilt and not to recognize herself as responsible. This phenomenon of dilution of responsibility has also been used on purpose. The archetype of such a usage is firing squads, in which the purpose of lining up soldiers to shoot dead a prisoner is nothing but to dilute and make invisible who gave the deadly shot.

In philosophy, Derek Parfit argued about such cases with his thought experiment about harmless torturers. His thought experiment goes: An individual is presented with a switch that, if turned, will increase a little the amount of shock a stranger is experiencing. The individual switches it on, and, as the stranger does not even seem to notice the slight change, leaves the room. But as hundreds of other people make the same decision, the victim is eventually screaming in pain. Parfit goes on concluding that: "Even if an act harms no one, this act may be wrong because it is one of a *set* of acts that *together* harm other people" (Parfit 1984, 70). In other words, an act may be wrong because it would harm other people if some conditions are fulfilled. In the context of environmental ethics, harm is not necessarily limited to human beings, but can include other-than-human living beings, ecosystems, other entities and milieus. What is remarkable in this formulation is that no specific group of individuals playing the role of harmless torturers has to be identified or treated as responsible as a group. Regardless of who belongs to the group, the individual action is wrong.

Cases of imperceptible yet cumulative impacts challenge the relation of causation between the agent's action and the consequence. In some cases limited in scope, it is still possible to identify a specific collection of individual agents who, each, did a similar seemingly insignificant action leading to a harmful consequence. This seems to be the case with Parfit's harmless torturers and with the firing squad. Yet, for most cases, it is hard to determine the scope of the collection of individuals involved, especially due to the interconnectivity of the social structure. In the example of the firing squad, should the officer who gave the order of firing also be included? What about the people who wrote the military codes following which the prisoner was sentenced to death? What about the society whose taxes are financing the work of the military? Many environmental problems fall under this category.

When discussing about the individual imprint on the milieu, we focused on the individual agent, and covered the range of traces and impacts this agent had on the world, regardless of her intentions and knowledge. Yet this perspective focused on the individual agent's imprint falls short when confronted with cumulative impacts. Instead of looking for an agent or a specific group as a scapegoat, it seems important to identify the source of harm first, and then to trace back causation. For most environmental problems, it is easier to identify the cause as a specific practice that is supported by a certain state of affairs. As we have seen in Chapter 2, practices are relatively stable and self-sustaining because of a loop effect between their symbolic reality and the material resources and structures they are based on. They are also formed in the context of a particular cultural imaginary.

The source of harm in many environmental problems is then a specific, usually common, practice that is repeatedly done by a collection of individuals, supported by other practices enacted by other individuals and formatted by the cultural imaginary. Practices are often mutually supportive, forming a network that I referred to as the social structure. Once a specific practice A (spreading herbicide) is found to be the direct cause of a particular harm (biodiversity loss), the other practices that are supporting practice A (buying products from intensive farming) can be identified as indirect causes significantly contributing to the continuation of that harm. Any individual that engaged in practices that either directly or indirectly contributed to harm can then be considered as contributorily responsible. To continue with the herbicide example, in societies where the food system is based on industrial farming heavily relying on the usage of herbicide, almost the whole population might be involved in relatively harmful practices to different degrees. We can describe this harmful impact (biodiversity loss) as a particular collective imprint. In Chapter 2, I defined collective imprints as the whole of individual and mediated imprints by a specific group or society. We have also seen that these collective imprints are historical, as they are formed and determined largely by historical processes spanning beyond the scope of an individual's life. These collective imprints on the milieu shape the milieu that itself is the medial matrix for individual agents. This circularity of the medial processes around the phenomenological agent and the community was discussed in Chapter 2. Now, where lies responsibility in the midst of these cycles?

Some authors have argued for collective responsibility, namely, that groups could be held responsible *as group* for some harmful consequences whose cause can be traced back to the whole group (Feinberg 1970; Corlett 2001). In other words, responsibility for collective imprints could be assigned to the group, regardless of the different contributions, or acts of protest of its members. Yet, I worry that holding group responsible for specific harm regardless of the differences in attitudes and actions among their members can lead to disastrous consequences, such as reinforcing discriminations and abusive power dynamics between members without accountability of the main individual perpetrators. Along this line, Mark Reiff argued that assigning collective responsibility to groups can also lead to "the escalation of violence and the watering down of moral strictures" (Reiff 2008; Smiley 2017). Indeed, assigning collective responsibility to the group disregards internal dynamics of oppression, domination and abuses, and may allow the few individuals who have committed the actions leading to the worse consequences to go away without being held personally responsible. Furthermore, members who feel unjustly held responsible for a harm that they almost did not contribute committing, or that they could not have prevented are likely to feel resentful and to reject responsibility. This can lead simultaneously to social unrest and the continuation of oppressive and abusive relationships within the group.

Because of these worrying possible consequences to the assignment of collective responsibility at the level of the group itself, I situate responsibility – and the question of ethics – at the level of individual imprints. The thorny question of the relationships between members of a group and of the relationship between members and the group will be addressed later. For now, let us focus on the question of responsibility of the phenomenological agent for her direct, mediated and domino-effect imprints on the world. By considering these three types of imprints, I intend to bypass the problem of cumulative effects. Indeed, because mediated imprints are so wide-encompassing (spanning up to imprints created by the very existence of the individual), they are highly likely to include cases in which the impacts of the actions of an isolated individual are seemingly insignificant, but when accumulated with others' actions, they produce harm.

Under an account including mediated imprints in the realm of an individual contributory responsibility, accumulated cases are covered. In Chapter 2, we discussed how the usage of herbicide in the field contributes to the global phenomenon of desertification. The farmer using the herbicide is also normalizing this practice, and simultaneously runs the risk of stigmatization by other members affected by the harmful consequences. The normalization of the practice of spreading herbicide is a mediated imprint that depends on other people's perceptions, understandings and behaviours. Buyers of vegetables produced using herbicides are also normalizing the practice and can be held responsible for this mediated imprint. As a whole, the collection of the different mediated imprints of an individual can be identified, and the individual can be held contributorily responsible for it. Then, considered jointly with the collection of imprints of other individuals, the total encompasses fully the whole of the cumulative impacts producing the harm. Similar reasoning applies to the aforementioned cases of climate change and online shaming. In Parfit's thought experiment about the harmless torturers, the action of turning the switch to increase the power of the shock is part of the direct individual imprint of the agent. Then, the violent pain experienced by the victim of the shock can be considered as part of the mediated imprints of the individual.

In sum, individual agents are contributorily responsible for their medial imprints, including their direct, domino-effect and mediated imprints. Yet depending on the level of harm and the distance in the causal chain between the agent and the harm, the contributory responsibility assigned to each agent for each imprint must vary in degrees. The severity of harm and its very existence must, of course, be taken into consideration. But as this harm is the first reason why we start looking for the source of causation of this harm and for the criteria to assign contributory responsibility, they cannot be the sole criteria determining the degree of responsibility.

4.1.3 Degrees of responsibility and distances between the agent and the harm

Consequences of one individual action can span at different geographical scales. As discussed in Chapter 2, environmental consequences of the usage of herbicide in one's field can affect not only the local environment (soil degradation in the field), but also environments that are distant but limited (pollution of underground waters), up to the global environment (spread of desertification). In turn, milieus and social structures can be affected by the consequences on the environment at different geographical scales. For example, the usage of herbicide may produce a loss of local knowledge of weeding, influence neighbours into using the herbicide and support the herbicide-producing industry.

Similarly, harmful consequences of a present action or of the medial imprint of a living individual have consequences at different temporal scales. Beside the immediate effect of killing weeds, the usage of herbicide can also contribute to inducing a drought next year through soil desertification. In ten years, the soil might also be dead and unusable, forcing migration. Even after the death of the farmer herself, her past usage of herbicide in her field may then lead her children to abandon the family house, even if, from the farmer's perspective, she devoted her whole life to the continuation of the familial tradition of farming and to the wealth of the family anchored in the ancestral family land. Finally, in some extreme cases, we can be almost certain nowadays that some of our actions are likely to keep having consequences on the environment even if the human species becomes extinct in some million years. The archetype of environmental consequences at such an extreme distance from us is the disposal of nuclear wastes.

Then, what is the relative distance separating the agent from the harm in the causal chain that might be used as criteria for determining the degree of contributory responsibility of an individual agent? This is not simply spatial nor temporal distances as described above. Indeed, if a gangster hides a bomb in the foundations of a shopping mall and sets it to explode after ten years, she would not be considered less responsible for the deaths caused by the bomb exploding, just because it occurred ten years after her action of posing the bomb. Similarly, a colonel pushing the button to launch a long-distance missile to destroy a city on the other side of the globe is not considered less responsible because of the spatial distance separating her from the targeted city.

Neither spatial nor temporal distances affect the contributory responsibility of the agent, which leads to the puzzling result that we might be contributorily responsible for harm happening after our own death. Because we attached contributory responsibility exclusively to the individual, this also implies that there is nobody to be held contributorily responsible for a specific harm if the agent is dead. In other words, contributory responsibility is not transmitted to descendants and to the next generations. Nevertheless, contributory responsibility can still be assigned to the dead agent and affect her reputation and eventually the projects she contributed to design during her life. This relates to the question of what kind of reparative actions are required once responsibility is assigned, and we will discuss that later.

I suggest that in the causal chain, the distance separating the agent from the harm refers to the quantity of interventions by other individual agents that is needed for the harm to occur after the agent's action. These interventions are the human-influenced events and conditions that affect the course of the causation after the agent's action. Other agents' actions and interventions can greatly influence the severity and the extent of the harm done. Attfield suggested that responsibility can be mediated through the actions of others, spatial and temporal distances, uncertainty and diffusion (Attfield 2009), but in the case of contributory responsibility, I use the word mediated in a meaning restricted to mediation by other agents. In my view, contributory responsibility is still direct despite spatial and temporal distances, uncertainty and diffusion. The contributory responsibility attached to the action of influencing another agent to commit harm will be of lesser degree than the one of the other, but not nullified.

For example, the loss of biodiversity in the field of the farmer who spreads herbicide is a direct consequence of her action, an impact of her action that is inevitable and does not require other people's interventions. The farmer thus has full contributory responsibility for the loss of biodiversity in her field. But full responsibility does not exclude other people's responsibility. The mayor of the village who personally advised the farmer to use herbicide on her field has a mediated contributory responsibility, that is of lesser degree than the farmer's contributory responsibility. Moreover, the members of the executive board of the company producing the herbicide will have a mediated contributory responsibility to a relatively small degree in the loss of biodiversity *in this specific field*. But as they will also be contributorily responsible for the loss of biodiversity in every area where their herbicide was spread, their total contributory responsibility for the production and distribution of the herbicide is likely to be important. Furthermore, when calibrated to the severity and the extent of the harm, the contributory responsibility of the executive board members is likely to outweigh the contributory responsibility of the farmer – but, of course, the objects of harm are different (in the case of the farmer, her own field, while in the case of the board members, the totality of the areas where the product was used), rendering the comparison doubtful.

4.2 Matrix and capacity responsibility

Up to now, I discussed only *contributory responsibility*, that is, the responsibility of an agent for contributing to causing some harmful consequences. In doing so, I temporarily put aside the various limitations influencing and limiting the agent in acting in a certain way. I focused on the imprint side of the cycle of codetermination of the milieu, and not on the matrix side. Yet, we cannot fairly assign responsibility to the individual for the whole of her imprints, including mediated imprints, without recognizing and taking into consideration her vulnerabilities and how her actions and omissions are shaped by the medial matrix. Ignorance, blindness and basic need constraints are all restricting the capacity of the agent to act responsibly. We must consider the common intuition that an agent cannot be held responsible if she did not have the capacity to act otherwise. In other words, contributory responsibility must be balanced with capacity responsibility.

An agent's capacity to do otherwise is not limited only by external co-ercion and lack of material resources. It begins at the very basic level of sense-making, at the level of how the agent makes sense of the situation. Then, it also questions what practical alternative courses of actions the agent could imagine and think of. Finally, of course, it also requires us to look for other constraints on the agent, from physical pain and mental health issues to financial limitations and accessibility to other resources.

Coercion and constraints on the individual's freedom of choosing alter-native ways of action overlap with the obstacles to environmental ethical actions that I divided into two categories. The first category of "psychologi-cal obstacles" refers to seemingly internal aspects that can be epistemological (such as absence of concepts, dogma) and emotional (such as laziness, apathy and lack of self-confidence). The second category which I referred to as "so-cial obstacles" covers more external constraints such as the lack of resources (financial, material, educational, etc.), institutional constraints (opposing laws) and social pressures and exclusion. They all affect the capacity of the agent to act responsibly.

I will first explore how we can conciliate responsibility with ignorance and psychological constraints. Then, I will discuss what responsibility agents have towards practices, and how to evaluate the easiness to change a behaviour in the assessment of the capacity of the agent to be responsible. Finally, I will discuss material constraints that limit the agents, such as the threshold of basic needs and her access to different resources.

4.2.1 Sense-making, ignorance and accessibility of counterfactuals

How we make sense of the world shapes the scenario and the scene in which we are making choices and taking actions. We understand the interconnected world that surrounds us from a particular subjective standpoint, borrowing concepts and conative guidelines to cultural imaginaries. The agent's capac-ity to act otherwise, a much-needed condition for holding an agent respon-sible according to common intuitions, is restricted by the agent's capacity to think in different ways. The latter is shaped by the medial matrix, through the process of participatory sense-making, inspired by elements of the cul-tural imaginary and guided and constrained by practices. In other words, the question of if the agent could or could not have acted in a different way is largely determined by the accessibility of counterfactuals to the phenome-nological agent at the time of taking the action. The capacity of the agent to think of other possibilities of action is influenced by the agent's imagination, experience, thinking habits, education, knowledge, etc.

The counterfactual capacity of the agent to think of ways to avoid doing a specific harmful action is limited and constrained by the state of knowledge of the agent at the specific moment of acting. In other words, the agent might be completely ignorant that her action, accumulated with others' actions, might have harmful consequences. Or the agent might be at the centre of social pressure preventing her to consider alternative pathways and pushing her to do the action. Moreover, the agent can be partially blinded by automatisms and by dogmatic worldviews imposing specific ways of thinking. What the agent can think of doing is also determined by her goals at the specific moment and by the social role she endorses.

So, not only the observer cannot penetrate the agent's phenomenological experience to assess her state of knowledge, but the agent herself might be confused and incoherent about her own knowledge and reasoning. Then, how are we supposed to assess the ignorance of an agent in order to dispense her from responsibility? Most authors resort to the conveniently broad idea of common sense. Yet, the amount of knowledge expected by this requirement of common sense varies greatly, especially when it comes to environmental problems. Because of the highly complex nature of most of the environmental systems, it is often impossible to claim full knowledge about the consequences of one's action. Moreover, the most accurate knowledge that most people have is not first-handed, but comes from learning and trusting information given by scientists and experts in a specific field, let it be ecology or climate sciences. This leaves most of us in a misty state of deciding what source of information is most trustworthy, and guessing out of probabilities what kind of actions we can take that will have the least harmful consequences.

The predictability of the consequences of one's action is a condition for the possibility to act otherwise. But this predictability is sometimes put into doubt. To the best understanding of a layman, debates about other scenarios of climate change depending on the amount of anthropogenic emissions predicted in the year are deeply obscure. It seems to be a common intuition that if the consequences of one's action are deliberate, namely, that they are not excluded from the agent's intentions of taking the action, then the agent is to be held fully responsible for them. This is impossible to verify from an observer perspective, but it remains crucially important for the phenomenological agent's perspective, as it influences self-assessment and mental health, through guilt and remorse.

Then, from the observer perspective, we are left with two possibilities. First, the consequences were predictable, yet ignored by the agent who deliberately went forward with her actions. Namely, knowledge about the consequences of doing a specific action was accessible and the agent was in a position to trust the information. In such a case, it seems reasonable to hold the agent at least partially responsible for the harmful consequences. Second, the consequences were unknowable for the agent who was left in a state of complete ignorance of the possibility of harmful consequences. Then, it is usually considered unreasonable to hold the agent fully responsible. Exceptions in which the agent is held responsible despite the consequences of her action being recognized unpredictable are usually the cases in which the agent took action while recognizing herself ignorant of some aspects, in other words, when the agent is aware of taking a certain risk in doing an action. In many legal systems, a person can be legally responsible for the consequences flowing from an activity even in the absence of fault or criminal intent, or even if the possibility of these consequences was unknown before the harm, which is referred to as strict liability (e.g. selling poisonous products). So, even if the harmful consequences were unpredictable, if the agent took risks involving others, it seems to be considered acceptable to hold her legally responsible for the harmful consequences.

What counts as accessible and trustworthy information? Compulsory education and common sense can help draw the lowest threshold of knowledge. Then, we can reasonably expect agents to seek more information about the possible consequences when they are about to undertake a new and exceptional action, because they are admittedly taking a possible risk. This can be under the form of experts' advice and Internet searches, but in any case, it involves other agents. With other agents come other potential blameworthy individuals. The first agent is responsible for seeking information about the possible consequences of her action from people who are socially recognized as experts in the field, that is, to check and beware of the source of information before trusting it. Other agents involved, such as people presenting themselves as experts, have a mediated contributory responsibility for the harmful consequences of the first agent.

To sum up, I argue that ignorance does not erase responsibility, but it does affect the degree to which we can reasonably hold an agent responsible. Considering only this aspect of ignorance to estimate the capacity of the agent to be responsible (which is not the only factor), we can already cross it with the different degree of contributory responsibility explored in the last section. If these consequences are deliberate, the agent should consider herself fully responsible not only for the direct consequences of her actions, but also for the domino effect and for the mediated imprints. Since the phenomenology of the agent is inaccessible to the observer, we cannot hold her responsible for all these without her own admission. Then, when the consequences are predictable yet ignored, it seems reasonable to hold the agent fully responsible for her direct imprints. As predictability of the consequences usually diminishes with the distance in the causal chain, it seems reasonable to consider the agent only partially responsible for the domino-effect consequences of her action. As mediated imprints involve other agents' decisions and actions, even if the consequences are predictable yet ignored, the agent can be considered partially responsible, because if other agents had acted otherwise, the harmful consequence might not have occurred. Finally, as in the case of selling poisonous products, it can be acceptable to consider the agent partially responsible for her direct imprints, because despite not having any blameworthy intentions, she still is the main causal trigger for the harmful consequence.

The reason why an agent can be held responsible for her action despite being ignorant about the harmful consequences lies in the risk she took when doing the action. When it comes to environmental problems, we are all in a state of relative ignorance regarding the complexity of the environmental systems. The idea of precautionary principle recognizes this ineluctable blindness. Its first globally accepted formulation appeared in 1992, in Principle 15 of the Rio Declaration:

> In order to protect the environment, the precautionary approach shall be widely applied by States according to their capabilities. Where there are threats of serious or irreversible damage, lack of full scientific certainty shall not be used as a reason for postponing cost-effective measures to prevent environmental degradation.

In the next years, variants of this formulation were included in national laws and international treaties all over the world. They all revolve around the idea that it is the responsibility of the agent to establish that her planned action is very unlikely to result in significant harm. The precautionary principle is usually intended for decision-makers, states, companies, etc. But it can be adapted for individual agents: Understanding their state of ignorance, individual agents must do what they can to prevent harm from occurring, namely, to research for more information and assess the risks before taking any unusual action.

How does this translate with lifestyles, habits and usual actions? As acceptable – namely, less risky – projects and actions at the political level change constantly with the progress of scientific knowledge, what lifestyles and habits are acceptable for the individual needs also to be updated with what they learn. Then, when recognized experts in a field are advising a radical change of lifestyle, individual agents are required to consider if the information is trustworthy, and what changes it implies for their own lifestyles. But no one can spend her time awake and energy exclusively to seek better information and knowledge about the possible consequences of her imprints. Crossing sources that have different interests and relevant knowledge is a time-consuming process. We are all subjects to confirmation biases, and we tend to search for, interpret and favour information in a way that confirms our pre-existing beliefs (Plous 1993). Confrontation with new pieces of information that conflicts with one's rock-bottom beliefs and challenges our worldviews and habits can be a painful experience. It might force individuals to reassess their worldview and threaten their self-image. Self-reflection is not a benign process. Beyond taking huge amounts of energy and time, it might also affect mental well-being. Mental health is a threshold that should not be crossed by crushing feelings of responsibility.

The precautionary principle as presented above also covers omissions, advising that uncertainty does not excuse inaction confronted to the risk of irreversible damages. In the account of contributory responsibility defended above, actions and omissions were treated symmetrically. Concerning lifestyle choices at the individual level, avoiding self-reflection and changes in one's lifestyle can be understood both as doing actions despite the predictability of harmful consequences, and omitting to change one's lifestyle. In any case, on top of being responsible for her imprints (including both actions and inactions), the individual agent can be held responsible for remaining ignorant, namely, for not taking any step to improve her capacity to be a responsible agent by acquiring relevant knowledge. So, the agent's capacity seems to be limited on both sides of the equation, by the precautionary principle and by the fragility of her mental well-being. On the one hand, the agent is supposed to constantly update her knowledge about the world in order to adapt her actions to minimize the harmful consequences of her imprint by following the precautionary principle. On the other hand, in order to remain mentally healthy and so, capable of continuing a fulfilling life, the agent must carefully deal with new pieces of information that might threaten her rock-bottom beliefs and affect her mental health.

In the wide-encompassing account of responsibility I propose, we are not only responsible for our actions and omissions, but also for our habits and lifestyles, that is, for who we become. In turn, who we are – that is, how we think, what we know, what we value, etc. – limits and constrains what alternative pathways are accessible to us. We might appear to be caught into circular reasoning. Not to give ignorance the power to erase responsibility is essential to circumvent this circularity. Indeed, if we have the access to new important knowledge and the physical and mental capacities to learn it, then we can be held responsible for our ignorance.

Nevertheless, we cannot assign equal responsibility to agents regardless of how hard or easy it is for them to leave "better" imprints. We have explored some psychological constraints such as the bowl of the cultural imaginary, the limitations of one's knowledge and the mental health of the agent. These constraints are not only internal to the agent, but they are largely influenced by the medial matrix. Lastly, these constraints are not exceptional such as brainwashing or mental impairment. They are totally normal and unescapable constraints on how any of us think, decide and make sense of the world. So, the medial matrix determines what we can think and imagine, but it also constrains and guides our behaviour through practices generally accepted in the milieu we live in.

4.2.2 Practices and easiness to change

Practices are largely affecting what we think as acceptable behaviours and actions, and what changes of behaviour appear to be doable. Once an alternative course of action or lifestyle is conceivable to the agent, practices of the particular milieu often put a spoke in the wheels of the agent's will to change towards a less harmful behaviour. Because practices are sites where the agent exercises her autonomy, they are also probed to normative judgements and ethical justifications. When the agent challenges, resists and opposes some specific practices, she exposes herself to social pressures, exclusion and isolation.

Setting aside contributory responsibility, we are concerned here with the evaluation of the capacity of the agent to be responsible, namely, to act responsibly and to be held responsible. Zooming out from what is accessible in thoughts to the agent to what concrete actions the agent could actually take, the capacity of the agent to be responsible depends on what alternative ways of action are available, and of the easiness of changing behaviour. The easiness to change the behaviour is related to the obstacles faced, and widely accepted practices often seem insurmountable. Then, in a world regulated by multiple normative intertwined practices, we also have to discuss what responsibility the agent has towards practices that she recognizes having harmful consequences.

After self-reflection or by learning new pieces of information about some harmful impacts of a particular practice, the agent might become aware that she can choose to follow an alternative practice from the one she used to. In other words, she realizes that she coordinates some of her individual actions with particular social norms, in a way that might be largely automatic as a result of internalized social meanings. Then she has the autonomy to deliberately decide to change her behaviour. She can do so by opting to follow an alternative accepted practice, by changing her behaviour slightly in ways that do not explicitly show her change of heart or by frontally challenging a practice and explicitly justifying it. We will explore these three options in relation with social obstacles such as the lack of resources (financial, material, educational, etc.), institutional constraints (opposing laws) and social pressures and exclusion.

Additionally, the agent is always situated in a particular milieu in which behaviours are organized by a social structure, that is, a specific network of mutually supporting practices. She might more or less freely play with the social structure. As a result, she might modify the social structure itself and improve not only her individual direct imprint, but also mediate progress in others' imprints. Such changes in the social structure are more likely to happen when the agent changes her habits openly, but this is not without important risks for herself.

The easiest way of changing habits without taking the social risk to disturb the established social structure is to change slightly one's behaviour without claiming it. Such changes in habitual actions of an isolated individual pass unnoticed by others and is highly unlikely to have significant mediate consequences on the social structure. It is the easiest option, because it faces only internal psychological obstacles, such as laziness and lack of self-confidence.

Now, let us take two examples of practices that have environmental harmful consequences to illustrate these options: the usage of private car linked to climate change, and the usage of herbicide leading to biodiversity loss and land degradation. For a household relying daily on the usage of a private car, the easiest change of behaviour would be minimizing the trips, use car-sharing or public transportation services on some days and opt for a more environmentally friendly vehicle. By doing so, they are not openly challenging an accepted practice, neither are they exposing themselves to social stigma. In the case of usage of herbicide, the farmer can reduce the quantity used to a strict minimum, shift to a less harmful product or use it only in extreme cases. Again, this behaviour is not frontally confronting the usage of herbicide, and it is unlikely to face criticisms by other farmers and villagers.

These recommendations are not new. Because they do not require drastic lifestyle changes and avoid frontal confrontation with the source of the harm, they are easily accepted by governments and enterprises and widely included in environmental education under famous key phrases such as "Reduce, reuse, refuse". But this voluntary approach is criticized by many for not being efficient enough to significantly reduce the harmful environmental consequences. Deliberately changing behaviour and opting to follow an alternative accepted practice can be necessary to stop contributing to harm. By following an alternative practice, the agent is also giving more weight to it in the social structure. Thus, it might influence the social structure in a lasting way, but still has limited mediated consequences on other agents sticking with the harmful practice. Moreover, when publicly changing of practice, the agent risks social isolation. Depending on the importance the agent and her social circle gave to the practice as part of their identity, the agent might lack social support, and struggle psychologically to construct a new identity fitting with her new behaviour.

On top of psychological obstacles and social threats, the agent might also lack resources to suitably change behaviour. In the case of the usage of a private car, the agent might choose to give her own car up, and to exclusively rely on car-sharing services, public transportations and alternative ways of transportations such as bicycle. But once the agent has identified a possible alternative, this alternative must still be accessible concretely, namely, the agent must have the resources to afford it. On top of the costs (financial, educational, etc.) of the new practice, the costs of changing are also to consider. These are blatant in the example of the herbicide. The farmer might choose alternative weeding technics, but it is only conceivable if she can access the knowledge and she can afford the tools. She might get educational and financial supports for the preservation of traditional knowledge by programs of the government. But for her, to make such a change in her practice amounts to risk her harvest, and so her whole income. Moreover, even if she does not openly criticize the usage of herbicide, she might face violent stigma from the neighbouring farmers who might reproach her not to clean her field properly and thus offering a breeding ground for contagious plant illnesses and insects harmful to the crops.

Finally, the agent can also opt to confront frontally the practice that has harmful consequences. Confrontation is hard and involves high risks for the agent. For the user of a private car, confrontation with other agents might be necessary to change the infrastructure to render public transportations more accessible. For the farmer, confrontation might involve explaining and educating fellow villagers, writing a letter to the company producing herbicide, asking politicians and environmental impact assessment agencies to take a step, etc. Public denunciation of the harmfulness of a collectively shared practice overlaps with activism and can face a range of obstacles depending on the context. On top of all the psychological and social obstacles mentioned before, the agent is likely to face severe lack of resources, especially when confronted with opposing laws, non-complying agents and companies, and institutions. In areas where law enforcement is lacking, agents might even face threats to their life and their relatives'. Sometimes, agents might also risk punishment from the state and persecution.

We have seen three different lines of actions that an agent can take when she concludes that the practice she was engaging into has harmful consequences. If we assume that the agent is more responsible for the harmful consequences of her practice when it is relatively easy for her to follow an alternative, we can then assign degrees of responsibility depending on the easiness to change as described above. Then, the agent could be held fully responsible for not changing behaviour slightly (first case), and partially responsible in the two other cases, depending on the risks she would have faced.

Now, what about the harmful practices of others that the agent is witnessing? Without engaging directly in a harmful practice, an agent might support it by not opposing it, and by supporting other agents engaging in it. If we include silence and looking away as supporting, then agents could be relatively responsible for evolutions of the social structure.

4.2.3 Vulnerabilities and powers

One might ask, then, should the mother burning charcoal every day to cook for her family, knowingly contributing to air pollution (and to a smaller extent to climate change), aware of the alternative possibility of using electrical stove and owning just enough assets to buy it be held partially responsible for air pollution? She is contributorily responsible, she is not ignorant of the harmful consequences and she has accessible alternatives that would be less harmful. But she might judge safer to keep her savings in case she has a medical emergency. I already suggested that one threshold of responsibility is mental health and well-being. Another baseline should be added: basic needs. Indeed, the purpose of assigning responsibility is not to freeze the agents in guilt, to fill them with anger and indignation or to starve them to death.

Thus, how much the agent is responsible must be calibrated with how much powers the agent has (Cripps 2013). After ignorance and accessibility of alternatives, and easiness to change related to practices, what powers the agent has is the last factor that influence the capacity to be responsible. On the one side of the spectrum, capacities start at the physiological level of basic needs such as food, shelter and sleep. On the other side of the spectrum, there are powers related to social role, financial wealth, skills, educational resources, etc. With great power comes great responsibility.

On the other hand, when an agent is deprived of the most basic necessities, she also loses large parts of her agency. For example, hunger and sleep deprivation diminishes human capacities to think clearly and to make decisions. Human needs are spanning from bodily necessities to mental health requirements. The idea of basic needs reach relative global consensus with its usage in the Millennium Development Goals of the United Nations. It was widely used to argue that states have an obligation to seek to fulfil basic needs of their population, and the international community must attend to the basic needs of the poorest. Different organizations include different definitions and details under the umbrella term of "basic needs" (e.g. Maslow 1943). Generally, the idea of "basic needs" is used in terms of human rights for their fulfilment, and not regarding the limits of individual responsibility.

Agency and vulnerability are two sides of the same coin, so an account of responsibility must take into consideration the vulnerability of the agent. The capacity of the agent to be responsible for harmful consequences on the environment depends on the fulfilment of the most basic needs. We cannot require an agent to avoid engaging in an environmentally harmful practice if the alternative practice does not allow her to fulfil her basic needs at least as much as the harmful practice. Thus, we cannot demand the mother burning charcoal to refrain from doing so if the buying of an electrical stove depleted some of her basic needs. I understand "basic needs" here as the very bottom of the pyramid, essentially breath, hunger, thirst and sleep. But if an electrical stove is offered to her by the government or a non-profit organization, and the electricity is affordable comparably to the charcoal, then the mother can be expected to opt for the latter option, less detrimental to the environment.

The same reasoning goes for the farmer using herbicide on her field. If the fulfilment of her basic needs depends on the limited yield of her field, then we cannot demand her to take the risk to jeopardize her source of income without having the insurance that the alternative farming practice will produce sufficient harvest to cover her basic needs, and the basic needs of the individuals dependent on her. But if she is given such an insurance by trustworthy knowledge holders, or if she is given the insurance that her basic needs will be fulfilled whatever the yield of her field, then she can morally be required to opt for the alternative farming practice deemed to be less harmful. Such insurances can be given by the scientific community, by states in the form of subsidies and poverty alleviation measures and by organizations that can be at the community level (villager's community, farmers' networks, etc.) up to a global scale (international human rights NGOs, developmental aid stakeholders, etc.). This brings us back to the general understanding of basic needs as rights, and to the role of members of the social structure. The mother burning charcoal is contributorily responsible for contributing to climate change, but wealthier bystanders are also responsible for not giving her the minimal conditions to enable her to change of practices. The individuals who have the possibility to empower other individuals are responsible to do so, and if they don't, their passivity can be treated as an omission and the harm resulting from it as part of their mediated imprint.

Kant famously placed the "should", that is, responsibility, prior to the "can", namely, the capacity (Kant 1999). This can be seen as a necessity given the universality of his categorical imperative. But to hold such a strong stance on responsibility is impossible in the wide-encompassing account of responsibility and contribution that I build here. As I argued previously, the "should" should not smother the "can". Based on the technological power obtained by humanity nowadays, Hans Jonas reversed the relation between power and responsibility established by Kant and argued that "you should because you can" (Jonas 1979, 230). He argues that not only responsibility depends on capacity, moreover, vulnerability (or the lack of capacity) responsibilizes the other capable agents. The responsibility parents have towards their infant is the archetype of this irrevocable responsibility. Because the infant is vulnerable, the parents – and the adults – are responsible to care for her.

But the picture of the matrix-imprint situation is far too complex to isolate capable perpetrators and vulnerable victims. Hardly any individual is fully deprived of her agency, at least not as long as she is able of thinking. But every individual's capacities are different. The mother burning charcoal to feed her family in extreme poverty is less capable to be responsible than the agent part of the middle class who uses her private car to commute to work every day, because the former has fewer powers to act differently and because her basic needs are at constant risk. Moreover, a high-ranking government official has more power than the commuter, because her social position entails her with it, simultaneously making her responsible for the consequences of it.

Therefore, we cannot assign equal responsibility to each individual agent who did an action contributing to harmful effects, regardless of the capacity of the agent, in the sense of capacity discussed here. Similarly, we cannot assign equal responsibility to each member of a group who contributed a harmful damage as a whole through intertwined causal relationships, regardless of their position in the group and of the internal power dynamics of the group. Every agent is entangled in webs of constraints that can be extenuating circumstances, starting with the lack of fulfilment of basic needs, up to social threats and lack of knowledge. Nevertheless, for the observer, social powers and basic needs provide a tool to estimate the capacity of the individual agent, without knowing all the details of her phenomenological situation.

4.3 Responsibility to do reparative actions

This account of moral individual responsibility for environmental harm is forward-looking. It connects a past harmful action or omission with a future expected actions. Because most of us contributed to causing environmental harm, be it solely by our existence, this past-oriented characteristic of responsibility matters less than its future-oriented converse. Past-oriented responsibility still matters crucially for individuals whom we can isolate as having played an active game-changer role in directly and significantly contributing to severe irreversible environmental harm. Yet, as the priority now is to mitigate harmful consequences, it seems to be more fruitful to focus on powers to bring about the necessary social changes than on attributing specific degrees of past-oriented responsibility to each individual. This account of moral individual responsibility for environmental harm is thus dynamic, indicating more ethical ways of life instead of targeting individuals with guilt and resentment.

This account of responsibility is also wide-encompassing regarding the possible objects of responsibility ascriptions, including actions, omissions, habits, lifestyles and one's very existence. We are responsible for having engaged into lifestyles that contributed to causing environmental harm. But as post-responsibility matters as much – if not more – as ante responsibility, we are crucially responsible for engaging into less harmful lifestyles *in the future*. The cycle of past actions and of critical assessment of their consequences leading to expectedly better future actions has no clear start and end – even if we could argue that it starts around birth and ends with death, as the agency is limited to living human beings.

For harm caused by past actions and omissions, I argued that who is responsible for what varies in degrees depending on the number of interventions by other agents in the causal chain, regardless of the temporal and spatial distances between the agent and the harm. I also argued that the agent's capacity to be responsible varies in degrees according to three main criteria: the state of knowledge of the agent, the social practices of the milieu and vulnerabilities and powers of the agent at the time of taking action. Finally, I added a basic threshold to clarify under what conditions responsibility is properly ascribed, namely, that the individual should not be exposed to (worse) threats to her basic needs by doing the environmentally better actions.

4.3.1 Reparative actions

Reparative actions are past oriented insofar as they attempt to make up for a harm committed in the past. In jurisprudence, reparation can be understood as replenishment of a previously inflicted loss by the perpetrator to the affected parties. Most of the time, it takes the form of monetary retribution. From a wider perspective on ethics and moralities, reparative actions can take diverse forms, often starting with recognition of one's contribution to the harm and apologies to the party harmed. Notably, reparation in law and ethics includes guarantees of non-repetition of the harmful actions. Now, under my wide-encompassing account of individual responsibility, to pinpoint a specific harm committed and to immobilize it in time can be complicated as lifestyles and habits are included as possible objects of responsibility ascriptions. Most of the time, responsibility is an ongoing and dynamic process. Then, what I refer to as "reparative actions" are not essentially to fix a past harm and to take actions to return to the previous state of affairs, but to changes one's behaviours in ways that reduce one's harmful imprint on the milieu. So, reparative actions can be prospective and adaptive depending on the standpoint of the observer.

In environmental law and policy, a well-known difficulty is to find immediate victims that can be directly causally linked to the harmful action, especially if what counts as a victim is restricted to living human beings. Yet, under my account, reparation does not require victims. Reparative actions are any actions taken to improve a certain state of affairs containing harmful practices, a state of affairs that one had contributed to bring about and sustain. Reparation is sometimes impossible, and often does not restore a previous state of affairs in its integrity. Some harm is irreversible, such as the extinction of a species. The literature on climate change is especially enlightening when discussing ongoing processual environmental harm. Policies addressing climate change are usually divided between mitigation and adaptation. Mitigation aims to greenhouse gas emissions and to stabilize the levels of heat-trapping in the atmosphere. Its goal is to limit the scope of the already occurring harm. Adaptation aims to develop technologies and to take actions to cope with the effects of climate change, namely, to adapt our milieus to the forecasted changes. Both mitigation and adaptation are necessary and included under "reparation", as they aim to improve an existing or expected state of affairs.

Then, what are reparative actions for individuals? The answer to this question echoes directly the diverse ways in which individuals have been shown to leave harmful imprints, while being able to act otherwise. Any changes in lifestyles and habits that contribute to improve one's imprint on the milieu can be considered reparative actions. Reparative actions have to be deliberate and well-intentioned. An individual living a less harmful lifestyle regretting her past high-consumption way of life cannot be considered as taking reparative actions, as she is not deliberately taking responsibility for her medial imprint. Moreover, members of the social structure have partial responsibility in designing a social structure grounded on the precautionary principle, and they have a responsibility in the design and transmission of a meaningful and non-abusive cultural imaginary, social structure and sense-making practices.

We discussed earlier three levels of easiness to change behaviour related to social acceptance. The easiest option of slightly changing one's behaviour is the most commonly advised because it seemingly applies to everyone regardless of their social status, culture or the political system they live in. Most of the time, these changes touch consumption habits, from boycott to recycling. If they are important steps in raising environmental awareness, they usually have only low effects and are not sufficient to bring the social changes necessary to prevent irreversible harm (Maniates 2001). Social scientists working on behavioural changes and environmental education have shown that appeals to change our attitudes and lifestyles to "advance a general 'human interest' are liable to be ineffective" (Kollmuss and Agyeman 2002, 247). Anja Kollmuss and Julian Agyeman explain this by the well-known gap between the possession of environmental awareness and displaying pro-environmental behaviours, showing that increasing environmental knowledge does not significantly trigger lifestyle changes. They argue that this gap does not appear because "'we' are irrational, but because the power to make a significant difference, one way or the other, to global or even local environmental change, is immensely unevenly distributed" (idem). If these findings point out to the need for common actions, they do not exclude individual actions, especially if we understand individuals as non-atomistic.

Setting aside questions of distribution of power that might have to do with distributive justice, we focus primarily on individual responsibility and actions. Changes in consumer choices are the most common kind of direct (in the sense of non-mediated) actions that can be expected from most individuals. Almost everyone is contributorily responsible for having consumed a product that could contribute to severe environmental harm, even if it seems insignificant such as a plastic bag. Moreover, the predictability of the consequences of consumption of potentially harmful products being high, unless the other options do not allow the agent to fulfil her basic needs, all the criteria are filled for the consumer agent to be fully responsible to change her consumption habits.

We can also create new meanings through participatory sense-making, that is, by interacting with other people. While it might be the less obvious aspect of reparative actions, it is nonetheless a crucial one. The creation and spread of ideas, meanings and stories that support and indicate the direction of sustainability as the desirable one is the cornerstone of environmental ethics. Through each interaction with other individuals, we can influence the process of participatory sense-making to foster the development of sustainability-prone meanings. Through time and exchanges, these meanings will furnish the cultural imaginary. They will present individuals with a wider range of positive options from where they will be able to pick up and build their own worldviews and identities. Finally, that is according to these worldviews and to the stories and meanings available that individuals will orient their actions and lifestyles. By actively engaging in participatory sense-making to develop meanings and stories encouraging sustainability, individuals might be able to shift what is considered "normal" and "acceptable" within their own milieu.

The purpose of ethical sense-making is to move the window of social acceptability towards sustainability and to make unsustainable practices fall out of it, that is, become socially unacceptable and politically not viable. The process of shifting the window of social acceptability is often slow and does not rest on one person's action. Yet, it is important for voices and ideas supporting sustainability to be heard and to be part of the cultural imaginary so that they can softly be relayed and shift the window of acceptable societal values and norms in the direction of sustainability. Notwithstanding, this slow and soft shifting of what is considered as "normal" within a particular milieu is especially significant as it restrains what project and policy are viable within this milieu. In other words, changes in the social structure are likely to be successful only if they are accompanied by a shift of this immaterial window of acceptability (e.g. Sardan 1995). No individual politician or small group of activists can efficiently push a particular policy forward if their political project falls outside of the window of social acceptability.

The imprints of the individuals sharing the milieu shift the window of what actions can be taken. Because of the lack of active involvement by the majority of agents in the design of the milieu, on the one hand, the milieu has a certain inertia to social changes in practices; on the other hand, the efforts of a few individuals to influence meanings and shift the window of social acceptability can have a comparatively important effect. We influence others by the way we speak about the world; by imagining and sharing ideas about sustainable alternatives; by engaging into practices differently, in slightly more sustainable ways; or by criticizing directly harmful practices. As individual phenomenological agents, we must design the medial matrix in a way that encourages sustainable practices and projects, according to the evaluation criteria discussed in Chapter 3. In their interactions with others, individuals can deliberately influence sense-making, practices and the cultural imaginary.

One-to-one exchanges are particularly powerful at transmitting meanings and influencing the worldview of another particular individual. In one-to-one exchanges, conversation partners are partially designing their own window of acceptability for their social interactions. It allows them to place it at a slightly different place than the usual limitations of what is socially acceptable within their milieu. Within one's own milieu, individuals have countless of opportunities for one-to-one exchanges through which they can reflect on their ideas and influence others' ideas. In particular, beyond influencing the landscape of norms of usages of the space, changing of practice publicly gives multiple opportunities for one-to-one exchanges.

More demanding pro-environmental actions include actions directed to change the social structure. Excluded from the realm of individual responsibility by extreme atomistic individualization theories, these actions are sometimes referred to as indirect or mediated (Kent 2009, 138). They range from voting and writing to government officials and companies managers, to actually taking more pro-environmental decisions related to management as a relatively powerful member of a group. They depend on the membership of a group (state, company, cultural or religious community, etc.), and on the political system or the internal organization of the group. Being a recognized member of a group usually gives some legitimacy to the agent to make some claims about structural decisions of the group. Moreover, membership also carries trust, which is essential for the agent's claims to be heard and given importance. I refer to these actions as "political", because they are related to the organization of the group, or the *polis*. They cover matters of political decisions that impact the whole group, namely, actions that are targeting governments and attempting to influence decisions by government officials and representatives, but also actions that target decision-making in companies, associations and any group who plays a role in designing the social structure. Political actions aim at changing a social structure or system that is co-created and maintained by a group of individuals with different opinions, roles and powers.

Last but not least, not all means are acceptable to bring an end to harmful practices and change the social structure, for the very reason that some means might trigger social unrest and undermine the sustainable efforts in the long term. Social stability can be considered as a necessary condition for environmental sustainability. Indeed, the development and the enforcement of efficient environmental laws can be assured only if the socio-political system is stable and flexible enough to allow adaptability. Crucially, stability here does not imply inertia, but flexibility and adaptability. To face appropriately environmental challenges, innovative lifestyles need to be imagined and new laws need to be drafted, while obsolete parts of the legal systems need to be amended or even abolished. A similar reasoning applies for infrastructures and material projects. Social stability is required for the long-term development and success of common projects such as the development of renewable energies.

This requirement for social stability as a necessary condition for environmental sustainability excludes some means of action. Social unrest and violent protests can jeopardize the success of collectively designed projects. Conversely, violent social movements are often triggered by other forms of structural violence within the social structure. Thus, to maximize the chances of success of sustainable projects, we need to aim for the ideal of absence of any form of violence in the socio-political system. In this sense, non-violence is a forward-looking requirement of any form of reparative actions, including civil disobedience (Sabl 2001). Indeed, violence tends to trigger emotional reaction and a crystallization of the opposing positions. It also engages the vicious cycle of retribution, excuses, justice and sometimes revenge. If a protest is to unleash long-term changes in a system seeking sustainable democracy, then it needs to be civil and non-violent, in order to foster dialogue instead of aggressive backlash.

Reparative actions must be civil, namely, they must be done in the respect for fellow humans, the absence of hatred speech or action, the absence of cruelty and the avoidance of violence and reckless endangerment of others (Milligan 2013). This basic safeguard also applies to any actions aiming at holding others responsible. Accusations are easily perceived as weapons of social violence. By frontally accusing others, one runs the risk of creating an agonistic dialogue (Singer 1994). An agonistic dialogue refers to the fact that interlocutors are not perceived as agent members of a dialogue anymore, but exclusively as adversaries. Because they may apparently reject the premises shared by the majority including the opponents, they risk creating echo chambers. They might receive increasing support from the audience already convinced, while ostracizing themselves more and more from the general public. Agonistic dialogue would rule out the possibility to take the common collective actions needed to achieve long-term sustainable changes of the social structure and the milieu.

This also rules out ecofascism. Ecofascism refers to the political model in which a totalitarian government "requires individuals to sacrifice their own interests to the well-being and glory of the 'land', understood as the splendid web of life, or the organic whole of nature, including peoples and their states" (Zimmerman 2008, 531–32). Theoretically, it could rely on militarism to defend the integrity of the "land". It is easy to see how this line of argumentation can easily muddle with nationalism. For example, before the emergence of the term, some critics already accused Watsuji Tetsurō's work of ecofascism (Parkes 1997). Hypothetically, ecofascism could be a tendency of governments to resort to extreme policies for the sake of environmental protection. Yet, the enforcement of such policies by ways of violence would almost inevitably trigger social protests and unrest that could, in the long run, hinder the success of the common sustainable projects. In short, the target of environmental reparative actions – sustainability – limits the means, that is, what kind of reparative actions can be legitimately taken by environmental activists. Overwhelming blame and ungrounded accusations are thus ruled out, as well as forcefully compelling individuals, because it would further deprive them from their agency instead of empowering them to take their responsibility. One cannot force another to be responsible or ethical as ethics and responsibility rest on internal decision-making. Nevertheless, it is sometimes necessary to intervene to prevent harm.

4.3.2 An interconnected world

At what scale should agents concentrate their efforts? Many environmental problems are caused by accumulated impacts and their consequences cross borders, such as climate change or some cases of transborder pollution. We need to pass moral judgements about issues of an unprecedented spatial and temporal reach. The relation between communities distant from each other is no longer optional. The need to discuss about global environmental ethics emerges as an effect of globalization. Scheuerman distinguishes several characteristics of globalization, among which two are especially relevant to assigning responsibility for environmental harm that spans across large distances (Scheuerman 2018). First, he points out the phenomenon of deterritorialization that comes from the ubiquity of large-distance transportation and communication. Deterritorialization challenges the common assumption of limiting the reach of ethics and responsibility to within given territorially bounded communities. Second, the increasing interconnectedness of systems and people led some to argue that there is no "global", but only interconnected locals, coining the word "glocal" (Wellman 2002). This line of argumentation seems especially fitting to the concept of milieu. As the other side of deterritorialization, interconnectedness also fosters the possibility to influence distant milieus.

Interconnectedness and deterritorialization raise the question of what happens with ethics beyond one's milieu and supposedly bounded community. They also highlight the issue of multiplicity of memberships, and the possibility to be part of communities that are seemingly a-spatial with cyber networking. Is deterritorialization and cyber networking threatening the relevance of the milieu as a matrix shaping individuals' worldview and behaviour? A milieu appears to be always local, rooted in subjective human existence. Reversely, different milieus are not necessarily spatially distant; for example, they might be different spaces used by distinct communities within the same city. The easiest case of individual shaped by her milieu is a sedentary person living in a bounded community rooted in a particular milieu. Then, community and milieu coincide, and the individual lives and acts informed, constrained and guided by the consensual significations and practices of this particular milieu.

Now, most individuals carry a multiplicity of milieus, such as the birthplace, travel places, workplaces and current living places. By travelling and moving, some individuals are influenced by other milieus. By meeting other travellers, the individual enters in contact with their respective significant milieus, for which one cares because they are part of the identity of the cherished others. By interacting with other distant communities through communication media, individuals are also influenced by their imaginaries and practices, and by the fate of their milieus. In short, yes, globalization exacerbates exchanges and contacts between different milieus. Nevertheless, the milieu where the individual exists and acts still plays a crucial role, because it shapes individual behaviours more than any other distant influences, by designing locally appropriate practices and usages of the space.

In brief, the individual is shaped by influences from multiple milieus and communities. And the individual is leaving traces, some harmful, in multiple milieus, including spatially and temporally distant milieus. Then, ethical responsibility *for distant harm* comes from the fact that the individual (even passively) contributes to leaving traces on this distant milieu. The individual responsibility for an environmental harm mainly depends on her causal contribution to the harm and on her capacity to have acted otherwise, that is, past-oriented responsibility. Conversely, forward-looking responsibility, namely, responsibility for taking reparative and adaptive actions depends on the present capacities of the agent.

Capacities are largely determined by the milieu. I distinguished three criteria affecting the capacity of the agent, each of them directly related to aspects of the milieu as a matrix (see Figure 4.3). The state of knowledge of

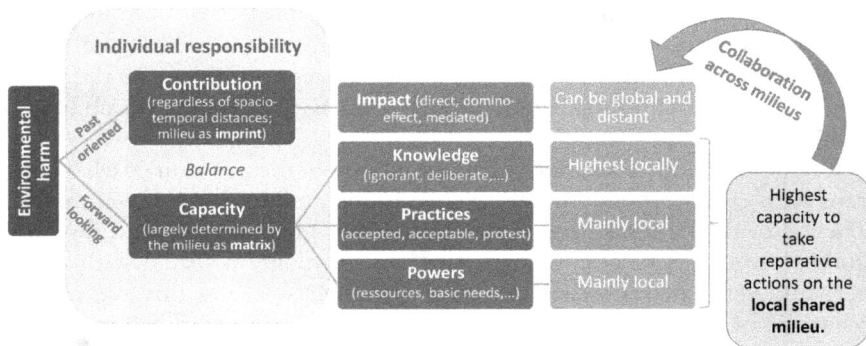

Figure 4.3 Responsibility for environmental harm and reparative actions

the agent and her capacity to imagine alternatives are mainly determined by how she makes sense of the world and the cultural imaginary she is in, both of which are features of the milieu as a matrix. The easiness for the agent to actually change her behaviour depends directly on the social acceptability and availability of alternative practices that are also part of the milieu as a matrix. Finally, powers and vulnerabilities of the agent are determined by the milieu she is in, and her position in her community.

As the milieu is local, so are most of the capacities of the agent. Generally, with the exception of scientific knowledge, the agent knows better her own milieu, so her capacity to find better solutions to environmental problems is higher in her own milieu. In other words, the knowledge of the agent is highest locally, so is her related capacity to imagine alternative practices. The agent also knows better the local practices than practices in distant milieus. She is best placed to estimate what other practices she could opt to without taking important risks. She can assess the social acceptability of some better practices, and the taboos attached to others. Usually, the agent is also better equipped to judge others' practices in her own milieu than in distant milieus, because they share similar conditions and cultural imaginary. When it comes to advising and criticizing others' practices, to be a member of the same community and to share largely overlapping milieus are crucial elements to support dialogue and to foster common changes. Generally, we are more prompt to trust and to accept advice from people who are close to us and who share similar lifestyles and struggles than from strangers. In short, it is easier for the agent to change her practices and to discuss of practices locally than to attempt to intervene in distant places.

The exception of scientific knowledge is critical. Indeed, the assessment of the extent of the human imprints on the environment is best done by sciences, that is, by minimizing as much as possible the interferences of human subjective and situated experience. Understanding what human practices lead to what severe consequences on environmental flows and systems necessitates scientific expertise, which requires collaboration between hundreds of scientists from different disciplines and from various milieus. Indeed, phenomenological agents entangled in the webs of relations of their own milieu might not be able to understand the roots of the environmental changes they witness. Of course, scientific expertise is oriented and tainted by the worldviews and projects of the scientists (Pascual 2013) and by the agenda of their source of founding. Nevertheless, scientific knowledge provides us with an epistemological depth much welcomed to assess environmental changes and to develop solutions to tackle what we judge to be environmental harm.

But scientific knowledge only provides information from which agents must conduct normative decision-making and actively change practices in a milieu. Most of these conative powers of the agent remain situated and anchored in the local milieu. Agents owning or having legal rights to use some strips of land have more power over its sustainable usage than on distant areas. Their practices on the land have direct consequences on it, and so they have full responsibility about it – unless they are constrained by others to use it in certain harmful ways. The agent is usually purchasing goods in local markets, at least for most commodities that are produced with high impacts on the environment, such as agricultural goods. Then, as a buyer, she also has the power to influence these local markets and to boycott commodities with the highest environmental impacts. Most of the social resources of the agent are also mainly local, such as reputation, power to influence others and right to vote. This does not exclude the fact that local actions can influence distant milieus, such as boycott. Still, the most powerful impacts that an agent can have are largely limited at the local scale.

In other words, the highest capacity of the agent lies in taking actions in the local shared milieu. Some changes necessitate common actions, which must be discussed, decided and coordinated with others, and the individual has a responsibility in engaging in fruitful dialogue to foster such common actions. Common actions can also be taken between individuals and groups from different milieus. A first axis is the sharing of information and ideas, without taking concrete common actions together simultaneously in several milieus, for example, thanks to agreements or partnership aiming at encouraging local sustainable programs. Exchanges that mainly take place on the symbolic level can encourage the groups to take common local actions within the local milieu, inspired by and under the glance of the public opinion of other milieus, and foster resources flow, be it information, knowledge, skills, technology or financial support. A second axis is collaboration between milieus that are already sharing some economic or cultural ties and so can build consensus and take actions *together*, that is, build common concrete projects to change practices in several milieus. This axis is more intrusive as it implies not only "sharing" of ideas and resources, but also a merging of leadership and project management. It might involve individuals making decisions regarding another milieu to which they do not belong, and thus raises the issue of interventionism.

4.3.3 Collaboration and common actions

To effectively tackle environmental problems, some changes require common actions or modifications of widely accepted practices, that is, deep reconsideration of some aspects of the social structure and of the collective infrastructures. Until now, I defended a forward-looking and wide-encompassing account of individual responsibility for environmental harm that varies in degrees depending on one's position in the causal chain, and on one's capacities to take reparative-adaptive actions. Yet, as members of a group, individuals are also perpetrators and complicit of the social structure of this group. I focused on individuals because they are the smallest unit of ethical agency. But they are composing and changing social structure. Then, what happens with responsibility for group actions?

A possibility would be to assign collective responsibility for past harm, that converts into collective responsibility to take reparative actions together as a group. This approach is used beyond the philosophical realm, in policy-making, law and environmental communication. Nations are often considered as groups who have a collective responsibility to reduce their greenhouse gas emissions, as established, for example, in the United Nations Framework Convention on Climate Change (Stone 2004). Companies are sometimes accused of being responsible for deforestation and biodiversity losses in some areas. Regardless of the different contributions, acts of protest of its members or internal power dynamics, groups seem to be treated as appropriate candidates for moral collective responsibility.

Four main problems appear when assigning collective responsibility to a group. First, it raises the problem of defining group's borders. Because of the closely interconnected ties of economy and trade and of the high mobility of the most powerful actors, to ground collective responsibility at the scale of the nation is doubtful. Second, collective responsibility erases dynamics of oppression and domination internal to the group. Some dominant individuals in the group who are the main perpetrators of the harm could continue oppressing the other individuals under the appearance of legitimacy because of collective responsibility. Third, it can lead to a dilution of responsibility and to nobody actually taking responsibility (Reiff 2008). Assigning collective responsibility to the group disregards internal dynamics of oppression, domination and abuses, and may allow the few individuals who have committed the actions leading to the worse consequences (e.g. decisions to allow the commercialization of some herbicides) to go away without being held personally responsible. Members of the group who gathered the main benefits of the harmful practices are then free riders at the expenses of other members of the group who had little to say in the continuation of these practices. Fourth, members who feel unjustly held responsible for harm that they almost did not contribute committing, or that they could not have prevented are likely to feel resentful and to reject responsibility as a whole. Such a situation can then lead simultaneously to social unrest, to the continuation of oppressive and abusive relationships inside the group and to the lack of effective reparative actions.

Political philosophers offered solutions to these problems. The solution to the first problem of sketching the border of the group collectively responsible was taken for granted for a long time as the nation-state (Miller 2004) or the society (Rawls 2005), which usually is taken to correspond with the nation-state. I showed that despite deterritorialization, most of the agent capacity still rests at the local scale, in her specific milieu. The specific area corresponding to this local milieu depends on the individual habitual mobility. The corresponding group who might be held collectively responsible for the milieu is then composed of the collection of individuals who are sharing the same milieu. The scope of the shared milieu will vary depending on the specific common actions that must be taken, ranging from the neighbourhood to the town, and maybe the region. But as usually powers diminish with distances, the focus of actions remains the local scale. Now, milieus are situated in specific nations, under legal systems and economical influences. Individuals are members of different groups and might have a say in the organization and actions of these groups. Consequently, another way of sketching the borders of a group is by looking at the *de facto* existence of these groups.

This means that the nation is not the only group individual agents are part of, and so it shall not be the only scale at which we assign and discuss collective responsibility. Notably, many nation-states do not offer any rights for individuals to get involved in decision-making leading to actions at the national level. In most representative democracies, this right is limited to choosing who will make these decisions by elections, instead of actually taking part in the decision-making processes. In addition to the group of the nation, individuals are parts of other groups such as the family or the village, but also groups of interests, as the company, regular buyers of some products, consumers of an electricity grid, clubs for different hobbies, religious community, etc. In any group which the individual is a recognized member, and which tolerates voicing of its members, then the individual can be expected to take actions to orientate the group's decisions and actions that might lead to environmentally harmful consequences.

Then, within a given group, three other problems may occur, namely, internal oppression, dilution of responsibility leading to passivity and resentment leading to social unrest. Young addresses these issues with her social connection model of responsibility, which "says that those who participate in the production and reproduction of structural processes with unjust consequences share a responsibility to organize collective action to transform those structures" (Young 2010, 184). Along this line, several authors argue that individuals have a duty to create institutions that encourage environmentally friendly practices (Caney 2005; Holly Lawford-Smith 2012; Tan 2015). For example, Cripps argues that "sets of individuals—as not-yet-organized collectivities or potential collectivities" can acquire duties to organize as necessary to "respond collectively to collective problems" such as climate change (Cripps 2013, 3). In addition, Lawford-Smith supports social-scaffolding for climate change-related decision-making because the complexity of environmental problems caused by habits pushes us to create and support policies and institutions to reduce the "number of moral choice-points in a day or make them significantly easier to resolve" (Lawford-Smith 2016). Aligning myself with them, I agree that, when possible, collaborating with others to change a harmful practice is essential and even takes priority over direct individual isolated actions (without erasing the responsibility to take them).

The variation in degrees of my account of responsibility allows to isolate individual perpetrators and to assign them different degrees of responsibility. This seems necessary, at least in the case of perpetrators who contributed in determining ways to causing severe environmental harm. A comprehensive model of responsibility for environmental harm should cover the ambiguous cases in which no clear causal link can be drawn and in which role-responsibility is ambiguous. In my model, different relations to the harm can be linked to different degrees of responsibility that demand different types of reparative actions.

Different types of reparative actions for harm generated by collective actions can be drawn from four types of relation to harm (Arendt 1987; Young 2010). First, those who directly causally contributed to the harm and who could have done otherwise could be considered "guilty". These "guilty ones" carry the highest degree of responsibility in my model, as they have a high degree of contributory responsibility and a high degree of capacity responsibility. They are expected to take important reparative actions and to be judged and punished. This seems defendable in the case of environmental harm too, especially when the individual played a causally determinant role in the occurrence of an environmental harm, and had the powers to prevent it, but did not. The question of enforcement is thorny, especially as the most powerful often plays beyond the national borders limiting the range of law. Nevertheless, it is crucial to be able to isolate these main guilty perpetrators in order to hold them accountable for their harmful actions. Holding them accountable means requiring them to take reparative actions to the whole extent of their capacities for the harm they have committed.

Second, there are the individuals who cooperated and were at least passively supportive of the enterprise of the guilty ones. In the case of environmental problems, these are the majority of the population. They might not have a social role carrying power to change importantly collective practices, and they might also think that these collective practices are not to be challenged, that they are "the world as it is". Individuals might, more or less consciously, reify social processes and practices so that it appears standing beyond the reach of their actions, and so beyond their responsibility. Beliefs in the reification of social processes often come along with passive support for those processes, even if they lead to visible harmful consequences. For some individuals, believing in the reification of social processes and practices is also a way of avoiding responsibility. As a result of balancing the low degree of contributory responsibility of the individual with the degree of capacity responsibility, the most vulnerable individuals might be exempt of blame because their degree of capacity responsibility is too low, and that they had not significant possibility to do otherwise without risking falling under the threshold of basic needs. In contrast, passive supporters of harmful practices who enjoy more power and resources must be held accountable for their passive support to the extent that they could have withdrawn it or even prevented or partially reduced the harm.

Third, individuals who are *de facto* member of a social structure which has some harmful impacts can take some action to distance themselves from these harmful practices. This third type of relation to harm can be considered as the minimal action to be taken by any member of a social structure towards the harmful practices accepted in that social structure. Indeed, withdrawal and detachment from a harmful practice come usually with low costs and low risks for the individual. As the individual takes action to detach herself from the harmful practice, she arguably shall not be blamed for it, as she withdrew her passive support to it. Off-grid and self-sufficient projects and communities are typical examples of this type of withdrawal.

Fourth are those who publicly resist and oppose some harmful practices, that is, political actions. If the individual is in a social position relatively powerful and that she is not risking to lose her means of subsistence and her own safety, then such preventive actions are required. Yet, resistance and opposition always should be orientated towards constructive solutions. Thus, publicly resisting a harmful practice must be accompanied with the development and support of alternative practices. Resistant actions depend on the political system the individuals are in, and on the risks involved. Moreover, resistance is a deeply individual and subjective action to take, so individuals who refrain from engaging in such activism shall not be blamed.

In sum, actions can be ranged from the ethically less desirable (being directly involved in causing environmental harm) to the ethically most desirable, namely, to support and develop alternative environmentally friendly projects and practices. To deal with any collective actions, it is necessary to analyse the details of the positions of each individual. This way, we avoid the problem of dilution of responsibility by assigning particular degrees of responsibility to each individual depending on her situation and capacity. By differentiating forward-looking responsibility and linking it directly to the individual situation, we likely also avoid the emergence of resentment coming with a seemingly unfair treatment. The absence of resentment lays the ground for taking common actions.

4.4 Pragmatic safeguards

What agents are responsible to do is determined by the balance of their contribution to causing the harm, and their capacity. Three main criteria were suggested to assess the capacity of an agent: predictability and state of knowledge of the agent (deliberate, predictable yet ignored and unknowable), easiness to act otherwise (slight behaviour change, alternative practice, public confrontation) and powers. In other words, the degree of responsibility of the agent decreases with the level of ignorance of the agent regarding both the possible consequences of her action and the possible alternative courses of action she might have chosen. The easier the alternative behaviour was for the agent, the more responsible she is for the harm she contributed to causing and for changing her behaviour. Finally, when it comes to environmental harm, an agent cannot be held responsible if the only alternative options she had to reduce the harmful consequences did not allow her to fulfil her basic needs, including mental health, at least as much as the harmful option she used. Conversely, bystanders to the situation of a vulnerable agent forced into committing harmful actions have a responsibility to empower the agent up to the point where she becomes able to act otherwise. If they omit to do so, then they have a contributory mediated responsibility for the harm produced by the action of the vulnerable agent. The responsibility of non-vulnerable agents increases along with their powers regarding a particular issue, that is, what one should do depends on what one can do.

These allow us to decide of a degree of capacity responsibility, to be balanced with a degree of contributory responsibility (Table 4.1). This is a complex enterprise and requires case-by-case analysis. Moreover, this is not enough to assess what the − relatively − powerful agents are expected to do to prevent a harmful consequence when they are not directly contributorily responsible for it − but might have a mediated contributory responsibility for it in virtue of being a member of the social structure.

Table 4.1 Degrees of responsibility

Degree of responsibility	*High*	*Partial*	*Low*
Contribution: distance in the causal chain	Direct	Domino effect	Mediated
Capacity: predictability	Deliberate	Predictable yet ignored	Unknowable
Capacity: easiness to act otherwise	Slightly change behaviour	Opt for an alternative practice	Public confrontation
Capacity: basic needs and powers	High powers	Basic needs fulfilled but no special power	Basic needs unfulfilled

We addressed three of the four worries regarding the assignment of individual responsibility for consequences of cumulated effects: We suggested that reparative actions should be focused on the local milieu and on the *de facto* groups the individual is a member of, and that dilution of responsibility leading to passivity could be avoided by isolating the main perpetrators. Further, the attribution to each individual of a particular degree of responsibility

Table 4.2 Tentative pragmatic safeguards when taking action

During our interactions with other people with whom we disagree

1 We must learn to **withdraw** to protect ourselves from threats to our basic needs, depression, burnout, etc.
2 We need to foster mutual understanding, agreement and common action by putting an **emphasis on common particular vulnerable experiences**.
3 The "accuser" should be **perceived as a fellow** by the "accused", so focusing on **local** actions seems to be the most promising strategy to hold others responsible.
4 We must take an **observer perspective** to assess the vulnerability of the other, namely, if our intervention is not harming her and causing her to become unable to fulfil her basic needs.
5 We need to understand what is our **position** and the one of the others within the political arena, because these positions can help us understand each other and what is at stake from different perspectives (while keeping in mind that these positions are flexible).

Regarding acting on others' milieus

6 We need to identify what attitude ourselves and our different conversation partners have regarding a specific practice to assess if our disagreements are **tolerable enough** to each other so that we can take common actions.
7 We must **prioritize communication** with members of other milieus over intervention, with the help of online networking to share information and ideas and put external public pressure for changes internal to the milieu.
8 We should **limit caring** to individuals to whom we are **strongly tied** (but can be spatially distant from us) and to their milieus.
9 Relations beyond our own local milieu must be **respectful**, that is, it rules out imposing unilaterally one's own project and policy – intervening in another milieu requires as much as possible **trust** and a certain extent of **consensus building** with the members of this milieu.
10 When the members of another milieu *refuse communication and continue engaging in practices that violate the necessary prohibition*, posing a threat to one's own milieu and survival, in last resort, **after open deliberation** within the groups affected and potentially intervening, the limits of collective intervention are the **protection of basic needs** and the consultation of the **scientific** consensus on the field.
11 The rationale for choices we make when intervening should be **transparent, accessible and acceptable** (or tolerable) for people from diverse milieus holding different worldviews, especially, if possible, for the people affected by the intervention.
12 We need to preserve **social stability** as a basic requirement for successful common sustainable changes.

regarding a specific issue could prevent individuals to feel resentment for being unfairly treated, and so avoid social unrest. The last worry of internal oppression remains, and it emerges especially when attempting to hold others responsible. To address it, we propose now some tentative pragmatic safeguards to avoid misinterpretations and misuses of this wide-encompassing account of responsibility. These pragmatic safeguards are listed in Table 4.2 and explored in the following sections.

4.4.1 Experiencing responsibility

The phenomenological agent who does her best to live ethically is like a tightrope walker on the thin line of ethical behaviour, between falling in the traps of abuses (doing too much and forcing others) or in the mist of ignorance (avoiding responsibility or doing not enough). She continuously must calibrate her ethical actions with regard to the vulnerabilities of the ones affected by her actions and her own vulnerabilities. From the perspective of the phenomenological agent, ethics is a self-limitation, that is, it involves refraining from engaging in practices and actions that are unethical. In the interactions, ethics become the limitation of what we impose on others.

We are building ourselves through interactions and relations. This process of becoming is never completed and lies at the heart of living. It also means that we can *never* escape our responsibility, as long as we are living (Smilansky 2013). So we should better make the best with what we have, and improve who we are, which appears to be the main purpose of individual human life once we accept the possibility of self-determination. As we are constantly in interactions, and the demands of the milieu are continuously slightly changing, we never can take for granted who we are and stop practicing the holistic virtues, namely, being curious (what we thought true before might have changed), adaptable (the habits that seemed to be the best before might not be so now), humble (we might have made mistakes) and self-reflective (examine our capacities and care for our vulnerabilities).

We cannot feel emotionally indifferent about these cycles and the challenging interactions and ethical demands of the milieu. My personal experience and my discussions with environmental activists testify for the emotional struggle that taking sustainable ethical action can be for the phenomenological agent. Drawing from her own experience and her work with activists, Sara Parkin describes this emotional struggle as the "sustainability emotional roller coaster" (in the left column of Table 4.3) (Parkin 2010, 22216). She divides her roller coaster between "positive feelings" (above the middle line) and "negative feelings".

Her account of negative feelings is, as she notes, consistent with the cycle of grieving by Elisabeth Kübler-Ross, which includes denial, anger, bargaining, depression and finally acceptance (Kübler-Ross 1969). Parkin describes

Table 4.3 Becoming: Emotional roller coaster, ethical decision-making and ethical action

Emotional roller coaster (Sara Parkin)	Ethical decision-making (my framework)	Obstacles to ethical action (pragmatic considerations)
↑ No looking back	**Taking responsibility**	**Social obstacles**
↑ Acceleration	↑ Take common action	Threats to life
↑ Collaboration	↑ Seek collaboration	Non-complying agents
↑ Collusion	↑ Humble	Institutions and state
↑ Positive deviance	↑ Adaptable	Opposing laws
↑ Curiosity and imagination	↑ Curious	Lack of resources (financial, skills)
↑ Point of no return	↑ Self-reflective decision	Social isolation, lack of support
↓ Renewed fear and anger	**Avoiding responsibility**	**Psychological obstacles**
↓ Naming and taming		Lack of self-confidence and fear
↓ Jim'll fix it	↓ Not my job	Laziness and apathy
↓ Fear	↓ Demands of immediacy	Lack of conceptual tools
↓ Denial	↓ No connection	Limited blindness, dogma
↓ Ignorance	↓ Reification	Absence of concepts

the loop of negative feelings as based in ignorance, that is, lack of awareness of sustainability as relevant to the individual. Once she acknowledges the existence of the problem, the phenomenological agent might try to deny it and her connection to it to avoid taking responsibility. She might be afraid and overwhelmed, and search for excuses to escape responsibility. Rejecting responsibility on someone else or on something else can respond to this emotional need, captured with expressions such as "not my job" or "Jim'll fix it". She might then try to bargain and minimize her responsibility and what she can do as much as possible – which Parkin refers to as "naming and taming". Finally, Parkin describes a renewed fear and anger, which eventually – hopefully – translates into indignation and allows the agent to escape the loop of negative feelings.

On the contrary to the cycles of grieving that relates to an unchangeable death, the emotional cycles of taking responsibility for sustainability can translate into positive indignation and empowered feelings when the agent realizes that she actually *can* do something about it. The agent starts the process of taking responsibility by taking self-reflective decisions regarding her own actions and beliefs. The positive feelings of Parkin's roller coaster are quite consistent with my account of taking responsibility for sustainability. Indeed, she mentions that the positive feelings of curiosity and imagination emerge as a result of crossing the "point of no return". After this point, Parkin argues that the agent becomes a "positive deviant", which she defines as "a person who does the right thing for sustainability, *despite* being surrounded by the wrong institutional structures, the wrong processes and stubbornly uncooperative people" (Parkin 2010, 1). She describes four "habits of thoughts" of the positive deviant: resilience (which corresponds to adaptability in my account), relationships, reflection and reverence (which relates to my description of humility as a holistic virtue).

Yet, individual positive deviance does not suffice, that is why Parkin includes as the final stages of her emotional roller coaster collusion (seeking encounter with supportive like-minded others, which amounts to seeking encounter in my account), collaboration (try and learn from each other, which amount to seek and build common ground in my account), acceleration ("do more" with more people, which amounts to take common action in my account) and "no looking back" (which might amount to the forward-looking aspect of my account of responsibility).

Parkin's emotional roller coaster is not a one-way road; it is cyclic, in the sense that individuals will "move up and down this roller coaster and may go around the loop or parts of it more than once" (idem, 216–17). Moreover, different individuals in different milieus might experience these different stages in various orders. We are constantly shattered and challenged emotionally by events happening to us and around us. We might even simultaneously deploy strategies to avoid taking responsibility about some aspects of sustainability, while taking full responsibility about others. While we present justifications tainted by our emotions for our actions, we are reinforcing some of our traits and habits, and (re-)defining who we are.

Taken in their continuity, these emotional cycles of reactions and actions amount to our lifestyles and the imprints that they produce. To a certain extent, we are also cultivating some emotions – which are not necessarily positive – and teaching ourselves how to deal with them and with external challenges. Crucially, the dynamicity of these cycles also indicates that there is nothing irremediable in wallowing ourselves in passive powerlessness. It also points out a way out of powerlessness into empowered engagement. In a nutshell, yes, we are vulnerable (not only to external and bodily circumstances, but also to our emotional roller coasters), but at the same time, we are agent, that is, we *can* decide how we act and change who we are and what we habitually do. Vulnerability goes hand in hand with agency, and the denial of one is as dangerous as the denial of the other.

From the perspective of ethics, every interaction is an opportunity for ethical action (leaving positive imprints on the world), and an opportunity to improve and build ourselves (receiving influences from the matrix). We live by these cycles of mutual influences with the world and others, and each interaction is a chance for us to use the space of choice of how we will act as a possibility to become who we want to be. In other words, while we are fundamentally becoming who we are in cycles of *codetermination* with the milieu (and others included in the milieu), we have a range of choice and freedom to *self*-determine who we are and who we become. Without this space for self-determination, we would lead deterministic existences in a deterministic world that would not leave any space for ethics, at least not for forward-looking ethics as individual responsibilities.

In any encounter, each interactor is negotiating her position in relation to the other and to the other's perception of herself. By and while doing so, they are also negotiating meanings, values and practices of the milieu and the social structure. These negotiations occur invariably in a context of latent conflicts between different positions. During the interaction, building dialogue amounts to finding a balance between accepting the difference of the other and seeking a common ground to develop an agreement. Along this line, ecofeminist theorists argue that difference must be respected and not homogenized by a process of assimilation, and that the interdependence and sameness at the ground of the difference must be recognized (Plumwood 1991, 2002). The unpredictability and the lack of control during the interaction reinforce tensions and vulnerability in any encounter. Welcoming the other while knowing that it might challenge some parts of yourself, your worldview and practices requires a certain courage and tolerance for error from myself and the others. The virtue of humility is rooted in this very intuition that any attempt to understand the other risks reducing the otherness of the other. The virtues of holistic conduct are tentative guidelines to foster positive interaction that involves being curious (to seek encounter with otherness to improve ourselves), adaptable (to be ready to change), humble (to accept our possibility for error) and self-reflective (to balance the new pieces of information with our temporary understanding of the situation).

We all have in common the fact that we are vulnerable. Vulnerabilities are intrinsic parts of what it is to be human, regardless of our particular situation and position in the social structures, and in time and space. Many moral considerations and ethical theories are rooted in the recognition of the vulnerability of the other, and, reciprocally, of our own vulnerability (e.g. Shue 2014), as demanding attention, care and respect. In the context of environmental uncertainties, the commonness of our human vulnerabilities can justify a globalized concern. In particular, Onora O'Neill argues that the persistence of human beings' vulnerability shouts for protection by means of justice. She adds that human beings may become "*deeply, variably, and selectively* vulnerable to the action of the particular others and the particular institutions on whom we come to depend for specific or often unavoidable purposes" (O'Neill 1996, 192; Luna 2006). These are two aspects of vulnerability: the fragility and finitude of human condition, and relational vulnerabilities that emerge as a result of interactions with other human beings, groups and institutions.

A conception of the self as continuously becoming, dynamically changing through cycles of codetermination with the milieu and others, sharpens these two intertwined aspects of vulnerability. As we are embodied beings, continuously adapting ourselves and thriving to improve who we are (or what we have), we are never invulnerable. We can be wounded physically and emotionally, and we are vulnerable to other people's ethical judgements, as well as to our own ethical self-evaluation. To a certain extent, we could even say that we are still vulnerable after our own death, as our reputation and memory become fully in the hands of other surviving people.

Because of our close interconnection and interdependence with vulnerable others (human beings, other living beings, species, milieus, etc.), any of our actions can threaten the very existence of these others. What are the limits and safeguards we should follow in our interactions with others when taking responsibility for our imprints? An intuitive way to search for these limits is to follow the adage "one person's freedom ends where another's begins", and to identify the reciprocal human vulnerabilities and remedy them. To nurture and shape milieus that provide individuals with the conditions for leading flourishing lives, we need to address the human vulnerabilities and to develop mechanisms to minimize them. Corinne Pelluchon interprets this aspect of ethics as a matter of "self-limitation" that questions "my right to be" and the place that I leave to others in my daily life (Pelluchon 2016, 304–5). Instead of framing the discussion in terms of competition for scarce resources with other beings, she proposes to draw principles of justice from the sharing of food and nourishment (Pelluchon 2015) and highlights the enjoyment we get from living and using together aspects of our environment (and, in my terms, of our milieu).

Thinking in terms of nourishment and enjoyment helps switching the perspective from the dark feelings coming along with responsibility and the realization that my imprints are harmful to others, to lighter insights into "who do I want to become", "how can I enjoy my life and sustain myself not at the expense of so much harm", etc. These more positive lines of self-reflection contribute to a sense of meaningfulness and proactive feelings of gratitude and engagement. In terms of the table above, they might help us to stay above the line of negative passivity and avoidance of responsibility. From such a positive perspective, the threat of the encounter becomes an opportunity for mutual enrichment. Further, the ever-changing and vulnerable aspect of the self turns into a skill for adaptivity. And responsibility goes from a constraining prison to a welcomed and meaningful guidance. This discussion brings us to a first pragmatic safeguard when taking responsibility for environmental harm, which is that we must learn to withdraw when necessary to protect ourselves from threats to our basic needs, but also from burnout, from depression and from the dark feelings of the sustainability emotional roller coaster.

4.4.2 Not enough: avoiding responsibility

When individuals keep taking actions that have harmful environmental consequences that are likely to affect the livelihood of others, while they have the full capacity to do otherwise, they must be held accountable. One may argue that the fact that the actions of the individual threaten others' milieus irreversibly gives legitimacy for these others to intervene and stop the perpetrator from continuing her harmful practice, but some limitations are needed to avoid oppression. The purpose of holding individuals responsible for their contribution to environmental harm is to convince individuals to change their behaviour and to empower them to do so. Then, the focus of holding others responsible is not moral blameworthiness, punishment or revenge. Instead, blame is appropriate "if and only if a reaction of this sort would likely lead to a desired change in the agent and/or her behaviour" (Talbert 2019). Responsibility can be used as a tool to clarify and enforce what the agent can be expected to do with respect to remedying or preventing harm.

Taking responsibility can be uncomfortable, and people use different strategies to avoid it. One is the reification of the social structure as way to avoid having responsibility for it, as it is an inert and unchangeable thing that has tremendous forces that one cannot possibly resist. Many philosophical traditions (e.g. several schools of Buddhism, Taoism and Stoicism) use the reification of natural and social forces as a tool to learn detachment and wisdom (Berger 1990; Moore 1995; Joseph 2014). Indeed, when one suffers from witnessing injustices and harms in the world, one has two options to end her own suffering. The first option is to address suffering in the world and to take actions to improve the situation "outside" of oneself. The other option is to change the self to transform the pain into seemingly more positive emotional states such as compassion, love or peace. To reach such a supposedly positive mental state, detachment from the hazards of the world is proposed as a first important step. A convenient way to foster detachment from the world's sufferings is to reify processes that bring about suffering into forces that cannot be stopped even if one wanted to. From a pragmatic perspective, trainings to improve mental health are indeed desirable, but they do not discharge the individual from her responsibility for the multiple harmful consequences she has on the world. Instead, inward exercises are important insofar as they foster a strong mental health needed to take positive actions in the world and to fight against harmful practices. For example, a society of peaceful, compassionate and happy monks travelling around in planes and eating food produced on the other side of the planet would not be sustainable and be the source of major environmental harm.

Besides reification, Young underlines three other ways used to avoid responsibility (Young 2010, 158–70). Some may claim that in the absence of direct and visible connection to other people, one cannot possibly be responsible for what happens to them. This denial of connection to others can also be heard from opponents of environmental policy changes, who sometimes argue that we have no relevant connection whatsoever to distant others, let alone other nonhuman beings. Because we have no connection, then we have no responsibility for their suffering, even if the latter comes as a domino effect of our actions. This excuse can be easily dismissed in my reasoning, as I address the anthropogenic environmental problems and take their causal connection to our lifestyles for granted.

When presented with scientific evidence regarding one's actual connection to some distant harm, one might argue that it is beyond one's energy and powers to care for them. Young refers to this as the "demands of immediacy", the fact that, absorbed by the multiple demands of our immediate relationship, we have no time or energy left to address wider general injustices (Young 2010, 161). In the context of environmental harm, responsibility can be overwhelming, as it covers potentially harmed living beings distant not only geographically, but also historically. Focusing on our responsibility for the structural processes of our milieu could help reducing the weight of the demands we cannot address.

The last way of avoiding responsibility that Young underlines is the "not my job" excuse, a cousin of the well-known "not in my backyard" mindset. The "not my job" excuse assumes the existence of the problem and the necessity to solve it. Individuals are able to claim that it is "not their job" just because the task of solving the problem is not assigned to anyone in particular. Yet, as Goodin writes, if it is recognized that someone has to do something about a harm, but as nobody is put in charged, then "we are all responsible for seeing to it that it be done" (Goodin 1995, 32). Environmental problems usually unfold in a multi-scaled manner that allows individuals to take actions to address them in multiple ways, in their different social roles. In countries where being a citizen ensures rights of criticizing and voicing opinion related to state policies and actions, citizens have a responsibility to check on the state activities, to actively support positive changes and to voice opposition to harmful practices.

The reification of harmful processes, the denial of connection, the demands of immediacy and the delegation to others' roles are all patterns of avoidance of responsibility that can be found mixed in individuals' argumentation. It is important to pay attention to these patterns insofar that they all echo some difficulties that individuals encounter when facing the idea of taking responsibility for the environmental harm that they contribute to commit. Reification patterns might express powerlessness and despair. The denial of connection possibly reflects feelings of insecurity or lack of feelings of commonness with what is excluded. Focusing on the demands of immediacy expresses the overwhelming emotions that might paralyze the individual. The delegation of responsibility to others might amount to criticize them for not taking their own responsibilities. Lastly, when discussing in groups, individuals often engage in blame-switching that comes without any constructive solutions, let alone remedial common actions.

When facing such a wide-encompassing idea of responsibility, it is common to feel powerless, insecure, overwhelmed and unfairly burdened. Some might find refuge in an identity as a passive victim. As a victim of an unbridled oppressive system or of the ill-willed conspiracy of a hidden powerful elite, the individual rejects the possibility of being a player in this game of designing social practices and the social structure. Self-victimization is especially hard to argue with when it is used as a deliberate technic to avoid responsibility. Over time, individuals might also build their own identity and self-image around the idea of themselves as more or less passive victims. As we have seen in the last chapter, individuals are building their identities in continuous cycles with their milieu. As a result of self-victimization, not only the world looks like terrifying webs of uncontrolled forces to which one cannot do anything but submit, but the individual might ostracize herself from the groups promoting proactive collaboration. In today's world, such a tendency is exacerbated by the wide accessibility to diverse information, letting the individual isolating herself in her self-confirmation bubble.

As agents standing outside of this self-victimization cycle, how can we reach to the individual and empower and encourage her to take her agency back, along with her responsibility? This question exceeds the scope of this book, but we can still present some basic safeguards. We need to make sure that our interlocutor would not end up unnecessarily harmed as a result of our accusation. Individuals who would have suffered a decrease of their quality of life threatening their basic needs if they had taken a less environmentally harmful action are exempt of blame for no taking it. This brings us to another important conclusion: Instead of accusation, blame or punishment, holding others responsible should be thought of in terms of collaborative common action. Proactive collaboration encapsulates a solution to the patterns of avoidance of responsibility discussed before. We need to establish a relation of trust with the other, insisting on the commonness of the issues and on the decisive role that she plays in the success of designing a more sustainable social structure. A second safeguard is to foster mutual understanding, agreement and eventually common action by putting an emphasis on common particular vulnerable experiences. This emphasis on common vulnerabilities can help establishing a relation of trust with the non-complying individual to avoid as much as possible agonistic dialogue, which could lead to a crystallization of each other positions at the opposite extreme, and seriously hinder the possibility of taking common actions (Essen 2017).

Trust often emerges from co-membership in some groups. As group members, we enjoy not only more trust than strangers, but we also have more tacit legitimacy to criticize the group's actions. A third safeguard is that the accuser should be, as much as possible, perceived as a fellow by the accused. Despite being part of the same group, the accuser still has an observer perspective onto the accused actions and failure to take her responsibility. As a fellow group member, the accuser could assess the capacities of the fellow members, as she is engaging with them in participatory sense-making, and she shares a similar bowl of cultural imaginary. As we have seen that the capacities of an individual are directly linked to her responsibilities to take reparative actions, a fellow member who knows the specific needs, vulnerabilities and powers of other group members is well placed to fairly estimate what changes of lifestyles could be expected from the uncooperative fellows. She is also likely to know best how to empower her fellows and how to take common action with them to improve their shared practices, without uselessly harming them.

Often group members live in closely shared milieus that become the main stages for negotiating common actions to improve them and reduce their environmental impacts. A group member may be able to build trust with her fellows and to trigger empowering moral shocks (Jasper 1997), as she also knows best the shared values and worldviews of the group. Finally, it is usually easier to take common action with fellow group members whose values and worldviews are well-known, than with strangers.

When group members fail to bring about change from within the group, they are also best placed to voice and communicate concern about the group practices beyond the fellow group members. Whistle-blowers are the typical example. They enjoy credibility in outsider's eyes precisely because they are expected to have better knowledge of the internal dynamics and difficulties within the group. Yet, group membership is dynamic, and whistle-blowers risk exclusion when they publicly criticize some internal practices. Indeed, other group members might have a defensive reaction and start an agonistic dialogue treating the whistle-blower and critics as enemies. These might lead to group fragmentation and conflicts of legitimacy within the group, especially when discussing the thorny question of repartition of responsibilities between members. Moreover, most of us are members of multiple groups that sometimes have conflicted interests. The individual might find herself divided between these multiple partially overlapping memberships that could be turned against her. These are all social risks to be balanced by the individual when choosing the best ways to promote changes in common practices. Direct accusation might not always be the most efficient way to push non-complying individuals to assume their responsibility.

Intervening to hold others responsible requires a twofold positioning. On the one hand, the accuser should take an observer perspective to analyse the situation and refrain from assuming that the other has identical capacities and vulnerabilities as oneself. It amounts to a fourth safeguard, namely, to be aware of and respect the difference in powers and needs of the other. On the other hand, the accuser should search what she shares with the non-complying individual and seek to appear as a fellow. Yet, membership claims are not always granted. Intervening in distant milieus and in groups that perceive the accuser as an ill-willed outsider can have drawbacks. In a world with multi-scaled borders and membership claims, the limits of a social structure are blurred. This can reinforce feelings of insecurity and powerlessness that could turn into anger and hate when facing accusations from a perceived outsider. These are risks to keep in mind seriously and balance to choose what way will be the most likely successful to make other individuals take their own responsibilities. A fifth safeguard can help, which is that we need to understand what is our position and the one of the others within the social structure, because these positions can help us understand each other and what is at stake from different perspectives – while keeping in mind that these positions are flexible (Lewicki, Gray, and Elliott 2002, 5–6).

4.4.3 Too much: interventionism, tolerance, respect and care

Until where and under which conditions are we justified to impose what we judge to be ethical on others, especially on other milieus covered with meanings and values that we do not share? We might be legitimate to intervene in others' projects, actions and ways of life if they threaten our own safety and flourishing. In particular, if they violate the necessary prohibition and if their actions (or omissions) are likely to lead to harmful consequences impacting our own milieus and ourselves, then we might be in a position to appeal to self-defence and impose some limitations on these "threatening" others. Yet, opening the door to such justifications is likely to worry more than one, as it might give way to imperialist or colonialist narratives of dangerous interventionism. Thus, we need to discuss the limits of responsibility, caring and interventions.

A certain tolerance to ideas and worldviews that contradict and oppose ours is necessary in the global context of pluralism of worldviews. The idea of tolerance distinguishes implicitly between three sets of beliefs and practices: the one we agree with, the one we find wrong but can still accept (objection yet acceptance) and the one we find wrong to the point we need to strictly reject them (rejection). The same practice can be evaluated very differently by different individuals. Still, some practices are widely considered intolerable and seem to require intervention and imposition of external ethical worldviews. An ethical system that tolerates everything would be void. Now, within my motivational framework, acts that violate the general prohibition – the irreversible destruction of global environmental systems – are deemed unethical and thus rejected and not tolerated. In other words, these acts require intervention. In turn, any action not violating the general prohibition but neither respecting the holistic virtues and grading low at the two criteria of evaluation (no harm and continuation) could be tolerable (3.3.2). Then, acting accordingly to the holistic virtues and the evaluation criteria is considered ethical. Finally, acting "ethically" but at the expense of one's health and basic needs (without hurting any relatives in the enterprise and still considering oneself happy and fulfilled) is supererogatory (Table 4.4). It is neither desirable, nor can it be deemed to be wrong. This nomenclature allows us to distinguish clearly the opinions of others and ourselves relatively to a practice, and then clarify the extent of the disagreements. Consequently, a sixth safeguard is that we need to identify what attitude ourselves and our different conversation partners have regarding a specific practice to assess if our disagreements are tolerable enough to each other so that we can take common actions.

As most environmental problems result from accumulated effects from consumption and production practices that are facilitated by other practices in a given social structure, we are all actors playing our parts in the negotiating game of what practices are socially acceptable and which are not. Some practices that are considered unacceptable in a specific milieu and sociocultural

Table 4.4 Tolerance: Supererogatory, ethical, tolerated and unethical practices

	The same practice A can be evaluated, by different individuals, as			
	Supererogatory	Ethical	Tolerated	Unethical
Definition	Too much to ask	Yes, we should do it	Not right, but still acceptable	Unacceptable, thus rejected
Reaction	Deserving praise	Neutral (good deed)	Neutral (bad deed)	Requiring blame and punishment
My framework	Act at the expense of one's health and basic needs	Act accordingly to the holistic virtues and the evaluation criteria	Any action not violating the general prohibition	Violation of the general prohibition
Policy (e.g.)	X	Projects respecting the two criteria (continuation, no harm)	Projects harmful at the short term with reparative strategies	Laissez-faire, exponential discounting

area are sanctioned by law. Laws are a central tool for groups to organize their practices. But laws have an historical inertia and can also be used to justify practices that actually are environmentally harmful. The relation between laws and environmental problems fully depends on the legal context attached to a specific geographical area limited by official juridical borders.

When we are confronted to an unethical practice – lawful or not, what kind of responsibility do we have for contributing to change and improve others' milieus? To limit harmful interventions in others' milieu, a seventh safeguard is that we must prioritize communication with members of other milieus over intervention, possibly with the help of online networking to share information and ideas and put external public pressure for changes internal to the milieu. Within the scope of the safeguards discussed here, transnational environmental civil disobedience could improve participation in the law-making process by giving a voice to beings that also endure the consequences of collective decisions while being under- or not represented, such as nonhuman living beings, ecosystems and future generations. In particular, some corporations and multinational enterprises span beyond borders and profit from legal loopholes, which leads to harmful environmental consequences. Facilitated by cross-border networks of environmental activists, transnational civil disobedience can be a key to tackle such cases by taking public opinion as witness and states as arbitrators (Droz 2019a).

Still, caring has limits, precisely because caring involves the power of deciding *what* to provide to fit the other *supposed* needs. When caring (in the strong sense of a serious consideration and provision for the needs of somebody or something (Misztal 2011)), we risk imposing our own needs and values (and the ones of our own milieu) to the others who, as a result, might suffer more than benefit from this care (Plumwood 2002). In the context of diversity of cultures, meanings and values, without clear and reciprocal communication, caring can then easily turn into domination and abusive dynamics. This does not mean that we should care exclusively for individuals who belong to the same local milieu. Many individuals develop friendships (characterized by knowledge of each other's needs and a strong relationship of mutual trust) with others who live in distant milieus. The care for these distant friends can be reported on their milieu as an important part of their cherished identity (Droz 2018). By extension, one *can* care for distant individuals and milieus, as long as this caring relationship is reciprocal and characterized by trust and communication. Without these precautionary conditions, care for distant individuals and milieus might be, in reality, an abusive relationship of the imposition of one's own worldviews and values on the other. As caring implies a strong emotional (and usually time) commitment, limiting caring to individuals to whom we are strongly tied and their milieus can be a relief. This gives us an eighth safeguard.

But the limits of caring are not the limits of responsibility. Responsibility goes beyond caring and changes form. While the discussion of responsibility to *care* fits the realm of the milieu, discussing responsibility in terms of *respect* for others and others' milieus seems to be suitable for relationships between milieus and beyond milieus. To behave respectfully might sometimes imply keeping our distances from the other whom or which we are respecting. Respect is distinct from care, for caring necessitates relatively strong ties. When interacting with non-intimate others in the aim of taking common actions to make some concrete sustainable changes in one's (or both) milieus, respect is fundamental. Respect in the process of taking an action that has consequences on a stranger and an unrelated milieu implies recognizing the potential differences in values, needs and functioning of the otherness. In other words, it implies recognizing one's own vulnerability and limited knowledge and legitimacy, including the possibility that one's best ethical action and behaviour might actually be refraining from intervening at all. Concretely speaking, respecting the members of another milieu rules out imposing unilaterally one's own project and policy on them. A ninth safeguard is that relations beyond our own local milieu must be respectful. In other words, intervening in another milieu requires as much as possible trust and a certain extent of consensus building with the members of this milieu.

What happens when the members of this other milieu refuse communication and continue engaging in practices that violate the necessary prohibition and pose a threat to one's own milieu and survival? In this last resort, calling upon rights is a basic consensual safeguard to significantly limit the ways of intervention. Any intervention at this scale is *never* a question of an isolated individual's decision-making. The decision and the ways to intervene are to be discussed and consensually agreed upon by the members of the milieus harmed by the irresponsible practices of the targeted group. For matters that have harmful consequences at the global level (such as climate change), such a consultation cannot be expected to be globally conducted with the world population. That is why we rest on the already agreed upon ideas and agreements such as human rights and international law and agreements. This has some limitations, similar to the ones of my working definition of sustainability that we discussed in Chapter 3 about the objection to self-determination, such as the fact that these "already agreed upon ideas" remain contestable and debatable.

Working with the idea of local milieu – partially – advantageously ditches the problem of defining borders. There are countless of milieus that lie on or across official national or regional borders. The concept of milieu primarily recentres responsibility and agency in the local grounds, from where they span along with the (potentially harmful) imprints of the individual. It also means that the official borders in no way erase the individual responsibility for her imprints that cross them. Neither does this *de facto* crossing of borders erases the borders. Instead, it extends them and creates new ones that impose some limits to the legitimacy of the individual actions and interventions beyond her own local milieu. From the perspective of the phenomenological agent, these limits are the limits of caring and the constraints of respect. In last resort, after open debates within the groups affected and potentially intervening, the limits of collective intervention are the protection of human rights and other wildly accepted international conventions, and the consultation of the scientific expertise on the field, including environmental impact assessments.

To consult scientific knowledge and expertise in a particular field before starting a project or intervening is a crucial safeguard not only for interventions across milieus, but also between milieus and within one's own milieu. Sciences are not unanimous, and they are at the centre of struggles for power and influences by different interests. Still, sciences provide us with tentative knowledge that is *the best certainty that we have* as they are, by definition, object to critical assessment and revisions. Social sciences are a central target of power struggles, as Bourdieu writes: "each specialist competes against not only other savants, but also professionals of symbolic production (writers, politicians, journalists) and, more widely, with all the social agents who, with very unequal symbolic forces and successes, work to impose their social worldview" (Bourdieu 1995, 106). Natural sciences – that play an essential role in environmental decision-making – are not exempt of these value-tainted influences, as we discussed in the context of biodiversity assessments (3.1.2 and 3.1.3). Still the scientific community provides us with widely debated and constantly revised information that informs us with the concrete aspects of practical decision-making. Scientific knowledge then crosses milieus and gives us a tool to assess and develop projects even on milieus to which we do not belong.

For all the reasons discussed above – from the risks of political and cultural domination to the fact that sciences are not absolutely value-free – to build an "evidence-based" sustainable project that seems to provide the best scientific solutions to a particular problem in a milieu *does not exempt* us from consulting the members of the local milieu and from renouncing to implement the project if it faces strong oppositions *during the dialogue.* But in the extreme case in which the community engaging in harmful practices refuses any form of communication and forces affected communities to intervene without any possibility for dialogue, relying on cross-fields scientific knowledge along with deliberative decision-making seems appropriate. In addition, for such deliberative decision-making to take place, scientific knowledge needs to be communicated in a way that allows non-scientific agents to see its potential biases and to take informed positions regarding it. As a result, the tenth safeguard reads: When the members of another milieu *refuse communication and continue engaging in practices that violate the necessary prohibition*, posing a threat to one's own milieu and survival, in last resort, after open deliberation within the groups affected and potentially intervening, the limits of collective intervention are the protection of basic needs and the consultation of the scientific advice on the field.

Before closing this discussion on interventionism on other milieus, a last point must be raised. Until here, we discussed interventions and relations with milieus that are geographically or spatially distinct from ours. We still need to address milieus that are *temporally* distant from ours. Indeed, the temporal borders of one's agency (one's limited life span) clash with the historical inertia of the consequences of the agent's actions that outlive her. Individual death does not sign the end of our imprint. It unfolds beyond. The evaluation criterion of continuation (3.2.1) incorporates transparently a way of thinking of one's actions and imprints as unfolding even after one's individual death. It encourages us to consider not only spatial limits to the legitimate exercise of our agency (and responsibility), but also temporal limits. Indeed, to leave imprints that have harmful consequences on the milieus shared by other individuals after our death appears to deserve some parallel considerations as leaving harmful imprints on geographically distant present milieus. Now, can a symmetrical reasoning that the one which applies to the harmfulness of imprints in the future also apply to the legitimacy of reparative actions that will also affect future individuals?

The distant future consequences of the imprints we leave in the present can be evaluated as harmful or not regardless of the identity or existence of the future individual agents and ecosystems. To the best of our current knowledge, if an imprint were judged harmful now, then it is to be avoided, following the precautionary principle. However, "reparative actions" aim at preventing harm as evaluated correspondingly to the present standards. We cannot reject the possibility that future standards of "harm" might differ from the present generally accepted ones, let alone from our own personal evaluation. But there are several ways in which we can mitigate the possibility of a large discrepancy between what is considered harmful (or good) now and what would be considered harmful (or good) in the future.

Because future people cannot protest our current decisions in the present, and because *who* will exist in the future *depends* on our current decisions (Parfit 1983, 1984, 2011), there seems to be an asymmetry of power between us and future generations (Usami 2011). Fabian Scholtes refers to this as "environmental domination", because our current idea of what should be sustained restricts and shapes other – in particular future – people's options and so exerts domination over others via the natural environment (Scholtes 2010, 290). Because we cannot dialogue with not-yet-born communities who will inherit the milieus that we are changing now by our imprints, all we can do is mitigate the risk of imposing on them imprints that they would suffer from and despise. Notably, this mitigation is not merely an act of charity towards completely powerless future people. Indeed, these future people have the monopoly of power over our posthumous reputation and on the success or failure of the projects we are dedicating our lives to. In other words, while we have the power to design what milieus we will leave to them, they have the power to destroy or to make flourish the milieus (as webs of meanings, values and projects) we cherished (Droz 2019b). In this sense, our responsibility and *what we care for* – the maintenance of the milieus we contributed to build through our lives – outlive us.

To mitigate temporal environmental domination, Scholtes argues that concepts of sustainability (which are, as we have seen in Chapter 3, the direction towards which orientate our actions) should meet three criteria: The (1) rationale for "choices we make about nature should be explicit and accessible" and should (2) "relate to a valuational reference that is acceptable to those affected", that is, concepts of sustainability should be (3) "aware and open to fundamentally different ideas of the good" (Scholtes 2010, 294–96). As we cannot discuss with the future inhabitants of the milieus we shape, to make the rationale for our decisions as accessible and transparent as possible (Scholtes' first criterion) seems to be a relevant minimal requirement. My model suggests a transparent rationale for choices characterized by the two criteria of evaluations (no harm and continuation), and insists on public deliberation and social interactions. Such a (ideal) decision-making process is likely to be understandable by future people, even if they disagree with some outcomes or some present values.

The possibility of future people disagreeing with the valuational reference used in our model is also mitigated by the fact that the two central criteria (no harm and continuation) used to evaluate projects have been historically and geographically widely approved. Moreover, as the definition of sustainability is formulated in terms of "conditions of possibility of continuation of self-determining" existences and entails the accessibility of diverse meanings and values and the enjoyment of an healthy global environment (without depending on other dominant human beings), it is reasonable to believe that it would be compatible not only with different worldviews held by currently living people, but also with the diverse worldviews of future generations. Finally, to remain open to a diversity of worldviews in the decision processes we currently make to choose how we shape our milieus lies at the heart of my model, as the virtues of holistic conduct show (curious, adaptable, humble, self-reflective, see subsection 3.2.3). By ensuring the accessibility of the rationale for choices, the acceptability of the valuational reference and the openness to a diversity of worldviews, we can mitigate the problem of temporal domination. This brings us to a penultimate safeguard, which is that the rationale for choices we make when intervening should be transparent, accessible and acceptable (or tolerable) for a people from diverse milieus holding different worldviews, especially, if possible, for the people affected by the intervention. Finally, a last safeguard was discussed previously regarding reparative actions: We need to preserve social stability as much as possible, as a basic requirement for successful common sustainable changes.

4.5 Summary

Within the framework of the milieu, individual *moral* responsibility for environmental harm results from the balance of contributory responsibility – the imprints on the milieu – and capacity responsibility – the influence of the milieu as a matrix. When it comes to environmental problems, we need to treat omissions symmetrically to actions, as omissions can have consequences to the same extent that actions do. Contributory responsibility covers all individual imprints, including habits, lifestyles and passive support to a state of affairs. This wide-encompassing account of contributory responsibility allows to tackle problems of harms generated by accumulated yet individually imperceptible effects, such as climate change. The proposal of addressing these problems as matters of collective responsibility of the group as a group was rejected because of fears about the possible implications on the repartition of responsibility internal to the group. My account of contributory responsibility is limited to individual contributory responsibility for her imprints, including the imprints that are mediated by the actions of other individuals. Contributory responsibility declines into different degrees, depending on the distance in the causal chain between the agent herself and the harm, that is, how many interventions by other agents were needed for the harmful consequence to occur after the agent's action. Conditions making the action actually possible prior to the agent actually doing it are excluded from this criterion, as they have to do with the capacity of the agent.

   Contributory responsibility for imprints on the milieu is to be balanced with capacity responsibility (see Table 4.1). The degree of capacity responsibility depends on the milieu as a matrix and decreases with the level of ignorance of the agent regarding consequences and alternatives, with the difficulty of behavioural change and with the vulnerabilities, but cannot go below the threshold of basic needs. This account of responsibility is forward-looking, that is, instead of questioning what the agent has done in the past (contributory responsibility), it focuses on what she could and should do (capacity responsibility) to prevent the occurrence or continuation of a harmful state of affairs, e.g. to leave less harmful imprints on her milieu.

Reparative actions are any deliberate and well-intentioned change in lifestyle and habits that contribute to improving one's imprint on the milieu. Positive imprints can help shifting the window of what is socially acceptable in a particular milieu, designing a social structure grounded on the precautionary principle and building meaningful and non-abusive cultural imaginary and sense-making practices. Pro-environmental actions directed to change the social structure should also be taken, but not all means are acceptable, as some means might trigger social unrest and undermine the sustainable efforts in the long term. Reparative actions should as much as possible preserve social stability, and extreme stances such as ecofascism are ruled out. Many changes towards sustainability require common actions that can be taken with people from different milieus, by sharing information, and take consensual actions together to change practices in several milieus. Yet, in our interconnected world, individual imprints can have consequences globally, but the highest capacity of the agent tends to remain anchored in the local shared milieu. Despite the need for common action, assigning collective responsibility to the members of a milieu *as a group* was dismissed because of the problems of defining borders, internal oppression, dilution of responsibility and resentment leading to social unrest. Indeed, within a group, different individuals have distinct responsibilities to take reparative actions aimed at influencing harmful shared practices, such as withdrawing one's passive support to them, or by becoming politically involved in resisting them, while supporting and building alternative environmentally friendly practices.

Our own vulnerability does not exempt us from taking responsibility for our imprints. Within the milieu, the agent can reflect on her ideas, influence others and launch coordinated actions through one-to-one exchanges in which she connects on a personal level with the other. The interactions she has with others echo the two movements of our interactions with the milieu as a matrix and as an imprint. In other words, the other is also a part of our milieu, and as such, she is also contributing to shape who we are (the movement of the matrix), and she is influenced by our interactions (the movement of the imprint we leave on the other). But these exchanges can also take a toll on the agent's well-being, and so one must also learn to withdraw to protect oneself. We are building who we are through these challenging ongoing interactions, through a dynamic process is hardly restful and tosses us around an emotional roller coaster between fearful denial and curious adaptable engagement, that is, between vulnerability and agency. These challenges can appear as opportunities for us to learn and improve ourselves, and we can reframe the discussion to ask ourselves what we enjoy, who we want to become and what place we give to others in our life.

Responsibility for environmental harm can be overwhelming, and individuals commonly use strategies to avoid taking responsibility, such as reifying the processes to set them out of their own agency, rejecting any kind of connection with these processes, arguing that they have "enough to do" with the ethical immediate demands, and delegating the role of solving the problem to others. Yet, the harmful imprints from individual avoiding their responsibility threaten the environment, our milieus, and even our lives and the ones of our children. Because of this, we need to hold irresponsible individuals accountable. To do this and to avoid misinterpretations and misuses of this wide-encompassing account of responsibility for environmental harm, twelve tentative pragmatic safeguards were added to guide agents in their journey to take ethical action (see list in Table 4.3).

To prevent others' harmful practices, we need to emphasize the shared experience of vulnerable fellow group members; at the same time, we must take an observer perspective to assess the vulnerability of the other herself, namely, if our intervention is not harming her and causing her to become unable to fulfil her basic needs. Care should be limited to close people whose needs we can understand and appropriately answer because of the relationship of trust and strong communication we share. Relationships with acquaintances and strangers should be characterized by respect. Imposing unilaterally one's own project and policy is ruled out; any intervention in another milieu requires trust and a certain extent of consensus building with the members of this milieu. The thorny question of interventionism emerges when others persist in engaging in practices that violate the necessary prohibition and cause harmful consequences on one's own milieu. When the members of another milieu *refuse communication and continue engaging in practices that violate the necessary prohibition*, in last resort, after open deliberation within the groups affected and potentially intervening, the limits of collective intervention are the protection of basic needs and the consultation of the scientific advice on the field. Finally, to mitigate the problem of environmental domination on future people, the rational for choices we make when intervening (including within our own milieu) should be transparent, accessible and acceptable (or tolerable) for people from diverse milieus holding different worldviews, as it might increase the chances that they are acceptable to the future people who will inherit our milieus after our deaths.

Note

1 Some aspects presented in this chapter were already mentioned in Droz (2020).

Bibliography

Arendt, Hannah. 1987. 'Collective Responsibility'. In *Amor Mundi: Explorations in the Faith and Thought of Hannah Arendt*, 43–50. Martinus Nijhoff Philosophy Library. Springer. https://doi.org/10.1007/978-94-009-3565-5.

Aristotle. 2014. *The Complete Works of Aristotle: The Revised Oxford Translation, One-Volume Digital Edition*. Edited by Jonathan Barnes. Princeton, NJ: Princeton University Press.

Attfield, Robin. 2009. 'Mediated Responsibilities, Global Warming, and the Scope of Ethics'. *Journal of Social Philosophy* 40 (2): 225–36. https://doi.org/10.1111/j.1467-9833.2009.01448.x.

Berger, Peter L. 1990. *The Sacred Canopy: Elements of a Sociological Theory of Religion.* New York: Anchor Books.

Bourdieu, Pierre. 1995. *La Cause de La Science, Comment l'histoire Sociale Des Sciences Sociales Peut Servir Le Progrès de Ces Sciences.* Paris: Actes de la recherche en sciences sociales.

Caney, Simon. 2005. 'Cosmopolitan Justice, Responsibility, and Global Climate Change'. *Leiden Journal of International Law* 18 (4): 747–75. https://doi.org/10.1017/S0922156505002992.

Corlett, J. Angelo. 2001. 'Collective Moral Responsibility'. *Journal of Social Philosophy* 32 (4): 573–84. https://doi.org/10.1111/0047-2786.00115.

Cripps, Elizabeth. 2013. *Climate Change and the Moral Agent: Individual Duties in and Interdependent World.* Oxford: Oxford University Press. https://doi.org/10.1093/acprof:oso/9780199665655.001.0001.

Davidson, Donald. 1980. *Essays on Actions and Events. Essays on Actions and Events.* Oxford: Oxford University Press. https://oxford.universitypressscholarship.com/view/10.1093/0199246270.001.0001/acprof-9780199246274.

Droz, Laÿna. 2018. 'Watsuji's Idea of the Self and the Problem of Spatial Distance in Environmental Ethics'. *European Journal of Japanese Philosophy: EJJP* 3: 145–68.

———. 2019a. 'Transnational Civil Disobedience as a Catalyst for Sustainable Democracy'. In *Rule of Law and Democracy – 12th Kobe Lecture IVR Congress Kyoto 2018*, edited by Hirohide Takikawa, 123–34. Stuttgart: ARSP.

———. 2019b, November 18. 'Tetsuro Watsuji's Milieu and Intergenerational Environmental Ethics'. *Environmental Ethics.* https://doi.org/10.5840/enviroethics20194114.

———. 2020. 'Environmental Individual Responsibility for Accumulated Consequences'. *Journal of Agricultural and Environmental Ethics* 33 (1): 111–25. https://doi.org/10.1007/s10806-019-09816-w.

Essen, Erica von. 2017. 'Environmental Disobedience and the Dialogic Dimensions of Dissent'. *Democratization* 24 (2): 305–24. https://doi.org/10.1080/13510347.2016.1185416.

Feinberg, Joel. 1970. *Doing and Deserving, Essays in the Theory of Responsibility.* Princeton, NJ: Princeton University Press.

Fischer, John, and Mark Ravizza. 1998. *Responsibility and Control: A Theory of Moral Responsibility.* Cambridge: Cambridge University Press. https://doi.org/10.1017/CBO9780511814594.

Golding, Martin P. 2005. 'Responsibility'. In *The Blackwell Guide to the Philosophy of Law and Legal Theory*, edited by Martin Golding and William Edmundson. Hoboken, NJ: Blackwell Publishing. https://doi.org/10.1002/9780470690116.

Goodin, Robert E. 1995. *Utilitarianism as a Public Philosophy.* Cambridge Studies in Philosophy and Public Policy. Cambridge: Cambridge University Press. https://doi.org/10.1017/CBO9780511625053.

Jasper, James M. 1997. *The Art of Moral Protest: Culture, Biography, and Creativity in Social Movements.* Chicago: University of Chicago Press.

Jonas, Hans. 1979. *Das Prinzip Verantwortung: Versuch Einer Ethik Für Die Technologische Zivilisation.* Frankfurt am Main: Insel Verlag.

Joseph, Harroff. 2014. 'A Daoist Critique of Reification'. In *Cross Currents: Comparative Responses to Global Interdependence*, edited by by Ian M. Sullivan and Cindy Scheopner, 97–116. Newcastle upon Tyne: Cambridge Scholar Publishing.

Kant, Immanuel. 1999. *Kant: Critique of Pure Reason*. Edited by Paul Guyer and Allen W. Wood. Cambridge: Cambridge University Press.

Kent, Jennifer. 2009. 'Individualized Responsibility: "If Climate Protection Becomes Everyone's Responsibility, Does It End up Being No-One's?"' *Cosmopolitan Civil Societies: An Interdisciplinary Journal* 1 (3): 132–49. https://doi.org/10.5130/ccs.v1i3.1081.

Kollmuss, Anja, and Julian Agyeman. 2002. 'Mind the Gap: Why Do People Act Environmentally and What Are the Barriers to pro-Environmental Behavior?' *Environmental Education Research* 8 (3): 239–60. https://doi.org/10.1080/13504620220145401.

Kübler-Ross, Elisabeth. 1969. *On Death and Dying*. New York: Scribner.

Lawford-Smith, Holly. 2012. 'The Feasibility of Collectives' Actions'. *Australasian Journal of Philosophy* 90 (3): 453–67. https://doi.org/10.1080/00048402.2011.594446.

———. 2016. 'Difference-Making and Individuals' Climate-Related Obligations'. In *Climate Justice in a Non-Ideal World*, edited by Clare Heyward and Dominic Roser. Oxford: Oxford University Press. https://doi.org/10.1093/acprof:oso/9780198744047.003.0004.

Lewicki, Roy, Barbara Gray, and Michael Elliott. 2002. *Making Sense of Intractable Environmental Conflicts: Frames and Cases*, 4th edition. Washington, DC: Island Press.

Luna, Florencia. 2006. *Bioethics and Vulnerability: A Latin American View*. Amsterdam: Rodopi.

Maniates, Michael F. 2001. 'Individualization: Plant a Tree, Buy a Bike, Save the World?' *Global Environmental Politics* 1 (3): 31–52. https://doi.org/10.1162/152638001316881395.

Maslow, Abraham. 1943. 'A theory of human motivation'. *Psychological Review* 50 (4): 370–96.

Miller, David. 2004. 'Holding Nations Responsible'. *Ethics* 114 (2): 240–68. https://doi.org/10.1086/379353.

Milligan, Tony. 2013. *Civil Disobedience: Protest, Justification and the Law*. New York: Bloomsbury Academic.

Misztal, Barbara. 2011. *The Challenges of Vulnerability: In Search of Strategies for a Less Vulnerable Social Life*. London: Palgrave Macmillan. https://doi.org/10.1057/9780230316690.

Moore, Robert J. 1995. 'Dereification in Zen Buddhism'. *Sociological Quarterly* 36 (4): 699–724.

O'Neill, Onora. 1996. *Towards Justice and Virtue: A Constructive Account of Practical Reasoning*. Cambridge: Cambridge University Press.

Parfit, Derek. 1983. 'Energy Policy and the Further Future Problem: The Identity Problem'. In *Energy and the Future*, edited by Douglas MacLean and Peter G. Brown, 166–79. Totowa, NJ: Rowman and Littlefield.

———. 1984. *Reasons and Persons*. Oxford: Oxford University Press.

———. 2011. 'The Unimportance of Identity'. In *The Oxford Handbook of the Self*, edited by Shaun Gallagher, 419–41. New York: Oxford University Press.

Parkes, Graham. 1997. 'The Putative Fascism of the Kyoto School and the Political Correctness of the Modern Academy'. *Philosophy East and West* 47 (3): 305–36. https://doi.org/10.2307/1399908.

Parkin, Sara. 2010. *The Positive Deviant: Sustainability Leadership in a Perverse World*. New York: Routledge.

Pascual, Unai and Eneko Garmendia.2013. 'A Justice Critique of Environmental Valuation for Ecosystem Governance. The Justices and Injustices of Ecosystem

Services'. In *The Justices and Injustices of Ecosystem Services*, edited by Thomas Sirkor, 161–86. New York: Routledge. https://doi.org/10.4324/9780203395288-25.

Pelluchon, Corine. 2015. *Les Nourritures. Philosophie du corps politique*. Paris: Le Seuil.

———. 2016. 'Taking Vulnerability Seriously: What Does It Change for Bioethics and Politics?' In 55:293–312. https://doi.org/10.1007/978-3-319-32693-1_13.

Plous, Scott. 1993. *The Psychology of Judgment and Decision Making*. Philadelphia: Temple University Press.

Plumwood, Val. 1991. 'Nature, Self, and Gender: Feminism, Environmental Philosophy, and the Critique of Rationalism'. *Hypatia* 6 (1): 3–27.

———. 2002. *Feminism and the Mastery of Nature*. New York: Routledge.

Rawls, John. 2005. *A Theory of Justice: Original Edition*. Cambridge, MA: Harvard University Press.

Reiff, Mark. 2008. 'Terrorism, Retribution, and Collective Responsibility'. *Social Theory and Practice* 34 (2), 209–242.

Sabl, Andrew. 2001. 'Looking Forward to Justice: Rawlsian Civil Disobedience and Its Non-Rawlsian Lessons'. *Journal of Political Philosophy* 9 (3): 307–30. https://doi.org/10.1111/1467-9760.00129.

Sardan, Jean-Pierre Olivier de. 1995. *Anthropologie et développement: essai en socio-anthropologie du changement social*. Paris: KARTHALA Editions.

Scheuerman, William. 2018, Winter. 'Globalization'. In *The Stanford Encyclopedia of Philosophy*, edited by Edward N. Zalta. Metaphysics Research Lab, Stanford University. https://plato.stanford.edu/archives/win2018/entries/globalization/.

Scholtes, Fabian. 2010. 'Whose Sustainability? Environmental Domination and Sen's Capability Approach'. *Oxford Development Studies* 38 (3): 289–307. https://doi.org/10.1080/13600818.2010.505683.

Shue, Henry. 2014. *Climate Justice: Vulnerability and Protection*. Oxford: Oxford University Press.

Singer, Peter. 1994. *Democracy and Disobedience*. New Jersey: Gregg Revivals.

Smilansky, Saul. 2013. 'Morally, Should We Prefer Never to Have Existed?' *Australasian Journal of Philosophy* 91 (4): 655–66. https://doi.org/10.1080/00048402.2013.775168.

Smiley, Marion. 2017. 'Collective Responsibility'. In *The Stanford Encyclopedia of Philosophy*, edited by Edward N. Zalta. Stanford: The Metaphysics Research Lab.

Stone, Christopher D. 2004. 'Common but Differentiated Responsibilities in International Law'. *American Journal of International Law* 98 (2): 276–301. https://doi.org/10.2307/3176729.

Strawson, Peter. 1962. 'Freedom and Resentment'. In *Proceedings of the British Academy*, edited by Watson Gary, 1–25. Oxford: Oxford University Press.

Talbert, Matthew. 2019, Winter. 'Moral Responsibility'. In *The Stanford Encyclopedia of Philosophy*, edited by Edward N. Zalta. Metaphysics Research Lab, Stanford University. https://plato.stanford.edu/archives/win2019/entries/moral-responsibility/.

Tan, Kok-Chor. 2015. 'Individual Duties of Climate Justice under Non-Ideal Conditions'. In *Climate Change and Justice*, edited by Jeremy Moss. Cambridge: Cambridge University Press. https://doi.org/10.1017/CBO9781316145340.008.

Usami, Makoto. 2011. 'Intergenerational Justice: Rights versus Fairness'. SSRN Scholarly Paper ID 2232607. Rochester, NY: Social Science Research Network. https://papers.ssrn.com/abstract=2232607.

Wellman, Barry. 2002. 'Little Boxes, Glocalization, and Networked Individualism'. In *Digital Cities II: Computational and Sociological Approaches*, edited by Makoto Tanabe, Peter van den Besselaar, and Toru Ishida, 10–25. Lecture Notes in Computer Science. Berlin, Heidelberg: Springer. https://doi.org/10.1007/3-540-45636-8_2.

Young, Iris M. 2010. *Responsibility for Justice*. Oxford: Oxford University Press.

Zimmerman, Michael E. 2008. 'Eco-Fascism'. In *Encyclopaedia of Religion and Nature*, edited by Bron Taylor, 531–32. London: Continuum International Publishing Group.

5   Conclusion

The general objective of this book was to build a motivational framework that supports sustainable behaviours and that can be consensual in the global context of pluralism of worldviews. The cornerstone of this framework is the concept of milieu, characterized by four aspects.

First, milieus are built through mutual relations and thus, the concept recognizes **the interdependence and inseparability** of human beings with their environment. We are shaped by the milieu as a matrix through participatory sense-making, borrowing meanings from the cultural imaginary and guided by practices; reversely, we are shaping the milieu by our imprints. In Chapter 2, I introduced the **conceptual framework of the milieu** to show how we develop worldviews and practices through mutual relations with others, with other species and nonhuman elements and with our milieu.

Second, milieus are **interconnected globally in a dynamic network**, which allows the concept to articulate different scales of analysis, such as the three levels of the individual, the milieu and the global spatiotemporal environment. But it also confronts us directly with the diversity of worldviews at the global scale, as milieus and individuals interact and mutually influence each other. It raises the question of how we could and should interact to foster sustainability.

Third, milieus are constantly changing and remain opened to sustainable changes. To guide individual responsibility and motivation, I provided a (re-) definition of **sustainability** in terms of milieu in Chapter 3. In terms of milieu, sustainability is the maintenance of the conditions of possibility of continuation of self-determining flourishing human existences. This temporal dimension gives a normative direction for sustainable ethical decision-making. It entailed a general prohibition, namely, that human beings should not individually and collectively deplete or destroy the global environmental systems to the point where these systems cannot independently provide healthy living conditions to current and future human beings. Preserving and developing the range of choice of individuals also entails cultivating meaningful, diverse and adaptable nurturing milieus. I further proposed two criteria of evaluation to facilitate the application of this approach to individual ethical decision-making, namely, minimizing harm and fostering the possibility of continuation in the long term.

Fourth, milieus are experienced by phenomenological agents, which links the milieu to the core motivation of the agent to adopt sustainable behaviours. Within human individual experience, agency and vulnerability are inseparable and concretely anchored within, and dependent on the milieu. In Chapter 4, I developed an account of **forward-looking individual moral responsibility** that results from the balance between capacity responsibility, which captures the individual vulnerabilities and needs shaped by the medial matrix, and contributory responsibility, which encompasses imprints that the individual's way of life leaves on the milieu. I suggested a series of safeguards to follow when taking action within our milieu and beyond our own milieu. These safeguards draw concrete limits within which we should take responsible action directed towards sustainability in collaboration with others.

The concept of milieu entails **four characteristics**: the external dimension of relational interdependence, the temporal dimension of dynamic changes, the internal dimension of phenomenological experience and the spatial dimension of global interconnection. These four characteristics allow the milieu to provide common grounds for a consensual ethics of sustainability. The concept of milieu is culturally sensitive and places ethics at the core of the relation between human phenomenological agents and their milieu. This book develops a conceptual framework of the milieu inspired by Watsuji Tetsurō's concept of *fūdo,* and applies it to the fields of applied environmental ethics and global political philosophy. By clarifying the imprint-matrix bivalence, the framework of the milieu highlights the inseparable intertwinement of relations binding individuals, communities and milieus. It could be used beyond the fields of philosophy and ethics, especially in interdisciplinary research about the relations between humans and nature that includes normative aspects. For example, the framework of the milieu could help to identify leverage points to bring transformative change towards sustainability from a systemic perspective (Meadows 1999; Abson et al. 2017), and to shed light on the "chains of leverage" (Fischer and Riechers 2019) connecting individuals, communities and their natural surroundings.

The **motivational framework** developed in this book from the keystone of the concepts of milieu and individual moral responsibility is directed towards sustainability. Sustainability is a precautionary multi-level dynamic process, which continuously redefines its objects. That is why the answer to the question "what should we maintain" is composed of "possibilities". My inclusive and flexible idea of sustainability attempts to be mindful of the global diversity of worldviews and open to further debates and improvements. For these debates to be fruitful, diverse and open, we need to keep the general prohibition and to minimize one-sided interventionism on others' milieus. The importance of confronting one's ideas with others and other milieus' worldviews is highlighted by a central consideration underlying this book, namely, the global scope of the problem and of the potential solutions to environmental changes. The flexibility of this conception of sustainability places changes in worldviews and milieus at the centre of the idea of sustainability as a dynamic ongoing process never exempt from reconsideration and discussions. Hence, sustainability remains an object of continuous negotiations, debates and redefinitions.

The motivational framework is also dynamic. It aims at guiding individual decision-making towards sustainability and at fostering collaboration with communities within and beyond the milieu. Ongoing individual decision-making processes are intertwined with and influenced by countless interactions with others and the milieu. The dynamic webs of relations that constitute milieus are changing as a result of individual decision-making, and of influences from other milieus, as well as environmental changes. Zooming out until the global scale, these dynamic webs of relations cover the globe and span historically from roots in the past to possible futures. They form a wide net intertwined with normative relations always attached to particular milieus, and they are perceived and deliberately revised by individual phenomenological agents. This wide net forming **global environmental ethics** is far from coherent, and is instead characterized by the multiple and highly diverse medial particularities and normative beliefs carried by different communities, covering milieus and embodied in individual ways of life. By virtue of the mutual interdependent relations binding together the individual human agents and their milieus, there cannot be one universal ethic of sustainability, but multiple ethical worldviews coloured and structured by the milieus in which they are lived.

Due to the **diversity of worldviews** and values that exist at the global level, conflicts regarding what practices are considered sustainable and what meanings and values are desired to be maintained inevitably arise. This is why an underlying goal of my book was to provide some practical keys regarding how to deal with these conflicts and take common actions. On the one hand, the minimalist normative premise that justifies sustainability for the sake of the preservation of flourishing human existences aims at fulfilling this goal by being widely consensual and acknowledging the diversity of worldviews and ways of living considered as a "good life". Some limitations of this argument and tentative solutions to them were provided in terms of objections to individual self-determination. On the other hand, crucial safeguards against one-sided interventionism and imposition of one's own values on others were discussed to prevent misinterpretations of my wide-encompassing account of individual responsibility. In particular, the nuances of tolerance were explored.

Zooming **in the experience** of the individual agent, I scrutinized the concrete ethical, practical and emotional implications of the abstract discussion regarding the milieu and sustainability. Confronted to the complexity of adopting and supporting sustainable ways of life, the phenomenological agent is likely to go through an emotional roller coaster, while navigating between social and psychological obstacles, the temptation to avoid responsibility and the motivation to take responsibility. To help the individual agent in this challenging journey, four **virtues of holistic conduct** were drafted: humility, adaptability, self-reflectivity and curiosity. First, given the limitations of her knowledge and capacities, and her ineluctable vulnerabilities, the agent was encouraged to seek humility. This humble precautionary stance acknowledges the relational interdependence that characterizes the framework of the milieu. Next, the second holistic virtue is adaptability, which reflects the goal of the motivational framework to encourage sustainable ways of life. Like the dynamic changing milieus, individuals can and should be ready to change their practices, habits and worldviews to direct their lives towards sustainability. Third, individual self-determination and flourishing were shown to be closely tied to ethical decision-making. The very idea of moral responsibility is bound to the virtue of self-reflectivity as the means to shed light on the internal experience of the agent and to make deliberate ethical choices. Fourth, inwards self-examination is not sufficient as agents cannot be taken out from the webs of relations that tie them to their milieu and to others. Thus, the final virtue of curiosity encourages individuals to look beyond their situated standpoints, which is a prerequisite for developing common grounds with others, and collaborate to build more sustainable social structures.

As soon as we deem ethics to be fundamentally relational – because we, human beings, are concretely relational – we closely connect it with politics in the sense of the art of living together. When we complete this perspective with the wide-encompassing account of individual responsibility, **ethics and politics** intertwine even more, as any action or omission of the individual reflects support or opposition to some dominant practices in the social structure. Within the closely interconnected webs of relations that are the milieu, any action of the individual, even the most mundane omission, becomes political as it encourages the continuation of some practices and discourages some other behaviours. In the political arena, we interact and meet with people we disagree with. What shall we do when we interact with individuals who reject our "common" premises and the need to change behaviours towards sustainability, and thus, who engage in lifestyles and behaviours that jeopardize the continuation of the existence of what we hold dear, be it our own lives, the ones of cherished others, our milieus or else. Until where does the apparent legitimacy that we get from defending our own survival and the one of our milieus go?

As much as possible we should foster dialogue and avoid violent conflicts, as efficient environmental strategies require common coordinated actions towards sustainability. In the motivational framework, sustainability is grounded on the milieu. To fulfil our responsibilities and render our milieus more sustainable, we need to build common grounds to foster collaboration with each other. The normative implications of our definition of sustainability, as well as our account of individual responsibility pertain to the *relations* one has with others and with ideas and projects. These relations constitute precisely the heart of ethics. A relation is composed of multiple interactions of many kinds, from face-to-face encounter with another human being to the discovery of an idea or a way of doing through reading a book or through observation of some elements of one's milieu. There are many ways of weaving a relationship, and sometimes the relation becomes harmful. Keeping in mind the criteria of minimizing harm, the question of how to develop "healthy" relationships emerges. I use here "healthy" as being the least harmful and potentially mutually enriching. In our conception of the self and the milieu, the webs of relations that the self is part of are constitutive of the self. In other words, the relation is not a link appearing between two already existing and independent elements, but it determines mutually these two elements. In this view, the loss of a relation is not simply the cutting off of one tie, but instead it might involve losing a part of oneself. The art to weave **healthy relationships** amounts to the art of seeking and developing common grounds.

What common grounds do we have to take actions with others towards sustainability? We share multiple **common grounds**, starting with the Planet Earth – the environmental systems supporting directly our existences – and our inescapable vulnerabilities as human beings. Planet Earth and the fertile soil on which we stand, and which supports the interconnected ecological systems provisioning nutrients and healthy conditions for human existence is a concrete common ground. We also need common grounds in the sense of reasons, justifications and shared ethical standards that bridge us across milieus and conflict to take common actions. The global environmental crisis highlights our vulnerability and fragility to changes in our environment. We have in common the facts that we value our existence and that we are vulnerable to these changes. These primordial **vulnerabilities** can give us a common reason to coordinate our beliefs and actions at least sufficiently to ensure sustainability. Such common grounds are far from granted, and vulnerability gives us only a starting point from which we can develop different justifications corresponding to different contexts and milieus. The milieu is a third common ground that encompasses the locally used areas and resources. These grounds include the local space carrying multiple possibilities for usages, and can be the stage of multiple conflicts of usages and appropriation.

The first physical sense of grounds as the **Planet Earth** is already a common ground in the literal sense, which can serve as a basis for developing common grounds in the second sense, that is, common justifications and reasons for actions. In this second sense, developing common grounds as common justification and motivation to take actions for sustainability is the overarching objective of this book. Finally, the third sense of common grounds reminds us of the importance of locating and contextualizing any process of developing such common justifications, as they will largely depend on the webs of meanings available in a particular milieu. What binds us, human beings, together globally are the intimate desires and everyday life concerns that make us human. To put an emphasis on common vulnerable experiences is a good starting point to deliberately develop common grounds across the specificities of the milieus and conflicting worldviews. In face of the urgency to make our milieus more sustainable, we can be expected to seek temporary consensus sufficient to take common action with others, and limit as much as possible harmful consequences. Consensus can include mutually respected tensions as potential sources of creation, motivation and stimulation for improvement.

This book roots ethics in a particular environmental milieu and sociocultural context. It proposes an account of the relation between the concrete local environment, sociocultural worldviews and individual decision-making and experience. As it clarifies the relations between scientific facts, values and sociocultural milieus, the motivational framework could have potential for the **science–policy interface** and policy-making. It may provide a tool to help to translate policies into norms that fit the local particularities of the milieu in terms of values, norms and beliefs. Moreover, this book connects ethics all the way from the scale of the internal experience of the individual agent up to the global reach and limits of the individual imprints, all mediated by the sociocultural milieus. As this book focuses on the individual human experience and individual moral responsibility, it comes as no surprise that more research is needed to explore how the flows of influences and ideas run between and across milieus. Borders are porous, be it boundaries of milieus, communities or embodied individual humans (Rees, Bosch, and Douglas 2018; Droz 2020). The individual appears as a nexus of multiple flows and influences that might originate far beyond from the scope of her understanding and reach. Investigating the **dynamics of these flows and processes** might be a promising alley for future research. Specifically, clarifying these dynamics might highlight the community's interactions with the medial matrix and the medial imprint.

Another exciting project would be to use the framework of the milieu to explore how other species approach and shape their world. While the motivational framework built in this book focused on the perspective of individual human beings as ethical agents, to consider other species as active partners and members of our communities that shape and are shaped by the milieu could be fascinating. Deliberately shifting *our* focus from human beings and considering human beings as members of multispecies communities could possibly encourage more sustainable ways of life. In other words, *thinking* that we are members of a **multispecies community**, adapting worldviews around this idea and *living as* members of complex biotic communities could lead us, human beings, to adopt more sustainable ways of life. The motivational framework could dialogue with more-than-human approaches that explore this possibility by shedding light on the close intertwinement of interdependent relationships in the biotic communities (Rupprecht et al. 2020), and by imagining sustainable futures in which humans are not giving themselves superiority over other species.

The motivational framework developed in this book does not exhaust the exploration of the normative aspects of sustainable solutions and the challenges of the current global environmental crisis. Still, it sketches possible avenues to help us understand how we relate to the natural world through our particular sociocultural lenses, and what we can do to foster more sustainable and flourishing ways of life. This can be done from our individual limited standpoint, as well as together with others. The motivational framework of the milieu does not ignore the challenges of the global pluralism of worldviews, of conflicts within the political arena or of how it can be overwhelming at times to take our own individual responsibility in the environmental crisis. The milieu connects us with others, with natural elements that are meaningful to us and on which we depend and with past and future generations. We cannot exist and live as humans outside of the milieu. As much as the milieu nurtures us, we have an ethical responsibility to cultivate milieus that are healthy and meaningful. Beyond being the very condition for flourishing human lives, the milieu provides us common grounds for global environmental ethics in an interconnected world.

Bibliography

Abson, David J., Joern Fischer, Julia Leventon, Jens Newig, Thomas Schomerus, Ulli Vilsmaier, Henrik von Wehrden, et al. 2017. 'Leverage Points for Sustainability Transformation'. *Ambio* 46 (1): 30–39. https://doi.org/10.1007/s13280-016-0800-y.

Droz, Laÿna. 2020. 'Living through Nature Capturing Interdependence and Impermanence in the Life Framework of Values'. *Journal of Philosophy of Life* 10 (1): 78–97.

Fischer, Joern, and Maraja Riechers. 2019. 'A Leverage Points Perspective on Sustainability'. *People and Nature* 1 (1): 115–20. https://doi.org/10.1002/pan3.13.

Meadows, Donella. 1999. 'Leverage Points: Places to Intervene in a System'. *The Sustainability Institute*.

Rees, Tobias, Thomas Bosch, and Angela E. Douglas. 2018. 'How the Microbiome Challenges Our Concept of Self'. *PLOS Biology* 16 (2): e2005358. https://doi.org/10.1371/journal.pbio.2005358.

Rupprecht, Christoph D. D., Joost Vervoort, Chris Berthelsen, Astrid Mangnus, Natalie Osborne, Kyle Thompson, Andrea Y. F. Urushima, et al. 2020. 'Multispecies Sustainability'. *Global Sustainability* 3. https://doi.org/10.1017/sus.2020.28.

Index

For Product Safety Concerns and Information please contact our EU
representative GPSR@taylorandfrancis.com
Taylor & Francis Verlag GmbH, Kaufingerstraße 24, 80331 München, Germany

www.ingramcontent.com/pod-product-compliance
Lightning Source LLC
Chambersburg PA
CBHW060300220326
41598CB00027B/4179

* 9 7 8 0 3 6 7 7 7 6 4 6 6 *